T0291823

CAMBRIDGE LIBRARY COLLECTION

Books of enduring scholarly value

Botany and Horticulture

Until the nineteenth century, the investigation of natural phenomena, plants and animals was considered either the preserve of elite scholars or a pastime for the leisured upper classes. As increasing academic rigour and systematisation was brought to the study of 'natural history', its sub-disciplines were adopted into university curricula, and learned societies (such as the Royal Horticultural Society, founded in 1804) were established to support research in these areas. A related development was strong enthusiasm for exotic garden plants, which resulted in plant collecting expeditions to every corner of the globe, sometimes with tragic consequences. This series includes accounts of some of those expeditions, detailed reference works on the flora of different regions, and practical advice for amateur and professional gardeners.

The Forest Trees of Britain

A keen collector and sketcher of plant specimens from an early age, the author, educator and clergyman Charles Alexander Johns (1811–74) gained recognition for his popular books on British plants, trees, birds and countryside walks. *Flowers of the Field* (1851), one of several works originally published by the Society for Promoting Christian Knowledge, is also reissued in this series. First published by the Society between 1847 and 1849, this two-volume botanical guide for amateur enthusiasts focuses on the trees found in British woodland. Johns describes each species, noting also pests and diseases, uses for the wood, and associated myths and legends. The work is noteworthy for its meticulous engravings of leaves, seeds and blossom, and of the trees in natural settings. Volume 1 (1847) provides an introduction to the botanical terms used. The species covered in this volume include oak, ash, beech and poplar.

Cambridge University Press has long been a pioneer in the reissuing of out-of-print titles from its own backlist, producing digital reprints of books that are still sought after by scholars and students but could not be reprinted economically using traditional technology. The Cambridge Library Collection extends this activity to a wider range of books which are still of importance to researchers and professionals, either for the source material they contain, or as landmarks in the history of their academic discipline.

Drawing from the world-renowned collections in the Cambridge University Library and other partner libraries, and guided by the advice of experts in each subject area, Cambridge University Press is using state-of-the-art scanning machines in its own Printing House to capture the content of each book selected for inclusion. The files are processed to give a consistently clear, crisp image, and the books finished to the high quality standard for which the Press is recognised around the world. The latest print-on-demand technology ensures that the books will remain available indefinitely, and that orders for single or multiple copies can quickly be supplied.

The Cambridge Library Collection brings back to life books of enduring scholarly value (including out-of-copyright works originally issued by other publishers) across a wide range of disciplines in the humanities and social sciences and in science and technology.

The Forest Trees of Britain

Volume 1

Charles Alexander Johns

CAMBRIDGE
UNIVERSITY PRESS

University Printing House, Cambridge, CB2 8BS, United Kingdom

Published in the United States of America by Cambridge University Press, New York

Cambridge University Press is part of the University of Cambridge.
It furthers the University's mission by disseminating knowledge in the pursuit of education, learning and research at the highest international levels of excellence.

www.cambridge.org
Information on this title: www.cambridge.org/9781108069144

This edition first published 1847
This digitally printed version 2014

ISBN 978-1-108-06914-4 Paperback

Selected botanical reference works available in the CAMBRIDGE LIBRARY COLLECTION

al-Shirazi, Noureddeen Mohammed Abdullah (compiler), translated by Francis Gladwin: *Ulfáz Udwiyeh, or the Materia Medica* (1793) [ISBN 9781108056090]

Arber, Agnes: *Herbals: Their Origin and Evolution* (1938) [ISBN 9781108016711]

Arber, Agnes: *Monocotyledons* (1925) [ISBN 9781108013208]

Arber, Agnes: *The Gramineae* (1934) [ISBN 9781108017312]

Arber, Agnes: *Water Plants* (1920) [ISBN 9781108017329]

Bower, F.O.: *The Ferns (Filicales)* (3 vols., 1923–8) [ISBN 9781108013192]

Candolle, Augustin Pyramus de, and Sprengel, Kurt: *Elements of the Philosophy of Plants* (1821) [ISBN 9781108037464]

Cheeseman, Thomas Frederick: *Manual of the New Zealand Flora* (2 vols., 1906) [ISBN 9781108037525]

Cockayne, Leonard: *The Vegetation of New Zealand* (1928) [ISBN 9781108032384]

Cunningham, Robert O.: *Notes on the Natural History of the Strait of Magellan and West Coast of Patagonia* (1871) [ISBN 9781108041850]

Gwynne-Vaughan, Helen: *Fungi* (1922) [ISBN 9781108013215]

Henslow, John Stevens: *A Catalogue of British Plants Arranged According to the Natural System* (1829) [ISBN 9781108061728]

Henslow, John Stevens: *A Dictionary of Botanical Terms* (1856) [ISBN 9781108001311]

Henslow, John Stevens: *Flora of Suffolk* (1860) [ISBN 9781108055673]

Henslow, John Stevens: *The Principles of Descriptive and Physiological Botany* (1835) [ISBN 9781108001861]

Hogg, Robert: *The British Pomology* (1851) [ISBN 9781108039444]

Hooker, Joseph Dalton, and Thomson, Thomas: *Flora Indica* (1855) [ISBN 9781108037495]

Hooker, Joseph Dalton: *Handbook of the New Zealand Flora* (2 vols., 1864–7) [ISBN 9781108030410]

Hooker, William Jackson: *Icones Plantarum* (10 vols., 1837–54) [ISBN 9781108039314]

Hooker, William Jackson: *Kew Gardens* (1858) [ISBN 9781108065450]

Jussieu, Adrien de, edited by J.H. Wilson: *The Elements of Botany* (1849) [ISBN 9781108037310]

Lindley, John: *Flora Medica* (1838) [ISBN 9781108038454]

Müller, Ferdinand von, edited by William Woolls: *Plants of New South Wales* (1885) [ISBN 9781108021050]

Oliver, Daniel: *First Book of Indian Botany* (1869) [ISBN 9781108055628]

Pearson, H.H.W., edited by A.C. Seward: *Gnetales* (1929) [ISBN 9781108013987]

Perring, Franklyn Hugh et al.: *A Flora of Cambridgeshire* (1964) [ISBN 9781108002400]

Sachs, Julius, edited and translated by Alfred Bennett, assisted by W.T. Thiselton Dyer: *A Text-Book of Botany* (1875) [ISBN 9781108038324]

Seward, A.C.: *Fossil Plants* (4 vols., 1898–1919) [ISBN 9781108015998]

Tansley, A.G.: *Types of British Vegetation* (1911) [ISBN 9781108045063]

Traill, Catherine Parr Strickland, illustrated by Agnes FitzGibbon Chamberlin: *Studies of Plant Life in Canada* (1885) [ISBN 9781108033756]

Tristram, Henry Baker: *The Fauna and Flora of Palestine* (1884) [ISBN 9781108042048]

Vogel, Theodore, edited by William Jackson Hooker: *Niger Flora* (1849) [ISBN 9781108030380]

West, G.S.: *Algae* (1916) [ISBN 9781108013222]

Woods, Joseph: *The Tourist's Flora* (1850) [ISBN 9781108062466]

For a complete list of titles in the Cambridge Library Collection please visit: **www.cambridge.org/features/CambridgeLibraryCollection/books.htm**

P. 68.

THE RUFAS STONE.

THE

FOREST TREES OF BRITAIN.

BY THE

REV. C. A. JOHNS, B.A., F.L.S.,

HEAD MASTER OF THE HELSTON GRAMMAR SCHOOL, CORNWALL;

AUTHOR OF "BOTANICAL RAMBLES."

PUBLISHED UNDER THE DIRECTION OF
THE COMMITTEE OF GENERAL LITERATURE AND EDUCATION,
APPOINTED BY THE SOCIETY FOR PROMOTING
CHRISTIAN KNOWLEDGE.

IN TWO VOLUMES.

VOLUME I.

LONDON:

PRINTED FOR

THE SOCIETY FOR PROMOTING CHRISTIAN KNOWLEDGE;

SOLD AT THE DEPOSITORY,

GREAT QUEEN STREET, LINCOLN'S INN FIELDS, AND 4, ROYAL EXCHANGE:

AND BY ALL BOOKSELLERS.

1847.

CONTENTS.

LIST OF ILLUSTRATIONS.

INTRODUCTION.

THE Author's object in preparing these little Volumes, is to furnish the lover of nature with such information respecting the trees which are either natives of Great Britain, or naturalised in it, as will tend to impart additional interest to his wanderings in the country. The reader, therefore, must not expect to find the announcement of any botanical discovery, any suggestions of new methods of planting, or recommendations for the improvement of timber. If he desires information on these points, he is referred to the numerous excellent works already in existence which treat on these subjects. But if he be merely desirous of exploring the wonders of nature as it is displayed in the more stately vegetable productions of his native country, it is hoped that he will find in the following pages, not, indeed, enough to satisfy his curiosity, but to stimulate him to fresh research. The Author assures him that even his own slender amount of scientific attainments can crowd the hedges and by-ways with countless miracles, which for the untrained eye have no being.

Scarcely any country in Europe is so favourable to the general study of the trees of temperate climates as England; for, without going so far as to assert that the number of native and introduced species exceeds that of all other states, it may be said with safety, that, whereas in most other countries the rare kinds are almost exclusively confined to botanic gardens and public institutions, the wealth and good taste of the English gentry procures for all trees worthy of introduction, and adapted to the climate, admission into the numerous parks with which the whole land is studded; where, without exception, for all purposes of observation and study, they are as much the property of the curious investigator, as of the lord of the soil himself. Scarcely a town in England is beyond a reasonable distance of some lordly demesne, abounding in fine specimens of most of our native trees, as well as many foreign ones, to the former of which the Author hopes to introduce his readers in the following pages.

Technical terms have been as far as possible avoided; but since, in describing the structure of a tree, it is necessary to apply to the several parts the conventional terms assigned to each part in scientific works, it has been judged advisable to give a general but slight sketch of the anatomical structure of a tree belonging to the class in which all the British trees are comprised.

THE elementary organs of all vegetables are either *cells* or *vessels*, which either singly or conjointly form what are called the *cellular tissue* and *vascular system* of plants. Cellular tissue is the simplest form of organised vegetable substance, and may be stated to be a combination of membranous cavities, the form and size of which are subject to endless variation even in different parts of the same plant. They contain watery juices, true sap, sugar, gum, resin, &c.: sometimes they appear empty, or are filled only with air. They vary in consistence, from the soft pulp of the peach, to the stony nut which it encloses. They have the power of transmitting fluid from one to another, though no pores have been to a certainty discovered in the membrane of which their sides are composed.

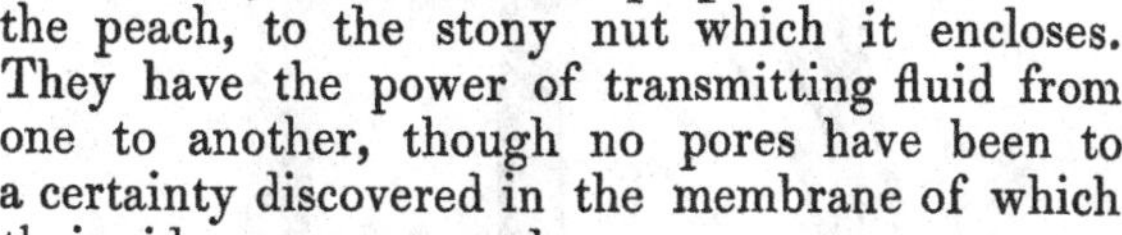

The vascular system comprises all those parts of plants which are known in common language by the names of fibres, nerves, veins, tubes, &c., all of which may be classed under the term "vessels." They are found in the root, trunk, branches, leaves, and flowers of trees, collected into bundles, but often so closely connected as to have the appearance of simple tubes. They are either of an uniform continuous substance throughout their whole length, when they constitute *woody fibre ;* or are composed of a thread, or collection of threads, twisted spirally so as to form a tube, and capable of being unrolled with elasticity. If a leaf of the strawberry, or young twig of the cornel, be broken, and the parts gently torn asunder, they will be discovered like fine cobwebs uniting the ragged edges. Under the microscope, if stretched, they resemble a corkscrew; but if examined at rest, their appearance may be compared to that of a bell-spring, that is, a coil of wire wound round a cylinder which has been afterwards removed. They are called *spiral vessels.* A third form of vessels is the *duct,*

which is a membranous, and not elastic tube, the sides of which are marked with transverse

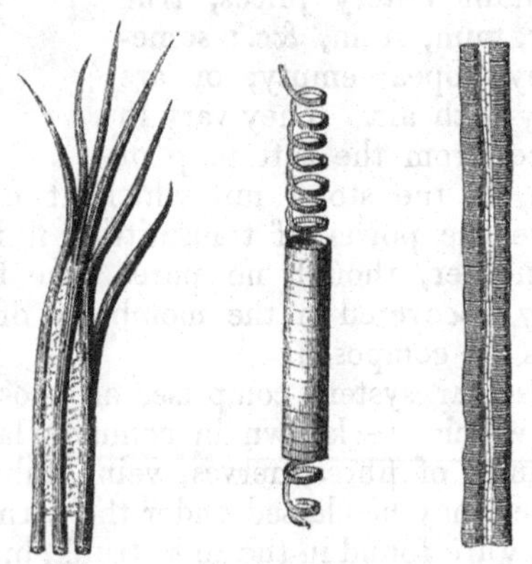

WOODY FIBRE. SPIRAL VESSEL. DUCT.

lines or spiral dots. These vary in form more than either of the last two, of one or other of which they appear, in many instances, to be modifications. The use of vessels generally appears to be to convey the ascending and descending currents of sap and air; but, owing to their extreme minuteness, and the difficulty of observation, there is much difference of opinion as to the exact purposes which they severally answer in the economy of plants.

Trees are either *endogenous* (growing inwardly), or *exogenous* (growing outwardly). All the trees described in these volumes belong to the latter class. A transverse section of the trunk of any one of these trees will present the appearance about to be described, varying more or less according to the species, age, and circumstances of growth of the specimen selected. The centre, called *pith*, consists of an uninterrupted column of cellular tissue, without any admixture of vessels. The cells are hexagonal: in their early stage they are green and filled with fluid; as the tree advances in age they become brown and dry, and sometimes hard. The pith never alters in size after its first year's growth, and is generally found to be larger

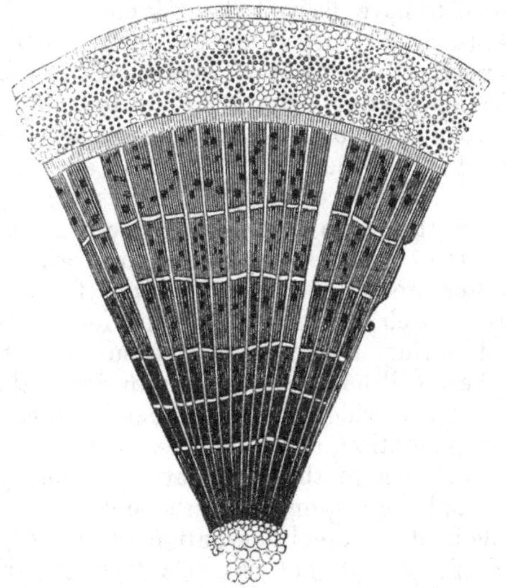

in strong lateral shoots than in the main stem, of which the Elder affords a marked example. It is enclosed in a thin tube composed of woody fibre and spiral vessels, called the *medullary sheath* (from *medulla*, marrow, to which substance pith was formerly supposed to be analogous). The medullary sheath retains its vitality after the pith itself has ceased to perform its functions, and extends its spiral vessels into the leaves, flowers, and fruit. Immediately without this is a dark-coloured cylinder, consisting of woody fibre and ducts cemented together, as it were, by cellular tissue. This layer was deposited during the first year in which the tree reached the height at which the section was made; it was then white, and is supposed to have derived its dark colour from the proper juices discharged by the ducts. The term *heart-wood* is employed to distinguish it after it has acquired this colour. The first layer of heart-wood is enclosed in a second, precisely similar, except that the medullary sheath is absent; and the whole substance of the trunk is made up of layers of the same form and structure, the exterior circles being softer and colourless, whence the outer wood is called *alburnum* (from *albus*, white). Each of these layers having been deposited during the growing season of a single year, there will be little difficulty in fixing the age of the tree, provided that it has not been checked by transplantation, and that it has been subjected to a climate where the termination of one year's growth and the beginning of the next have been well defined by the intervention of winter, as a period of rest. In hot climates trees experience no season of perfect cessation from growth; and when this is the case, the annual layers are so confused, as to permit no certain criterion of age. The same may be said of trees of great antiquity, for in these also the annual circles are often imperfectly defined. The external coating of the tree, the *bark*, is composed of two distinct parts. The outer, called the *cuticle* or *skin*, consists entirely of cellular tissue; in young stems it is more or less succulent, but in old stems becomes withered and dry. Underneath the cuticle is a layer of woody fibre, called the *liber*, or *inner bark*. This is succeeded by another layer of cuticle and liber, there being as many rings of bark as there are of wood. As each of these rings is deposited annually inside the old bark, the outer one of all was formed during the same year with that circle of wood which is next to the pith; but since the circumference of the tree when the first layer was formed was much smaller than at any subsequent period, and its power of expanding was very limited after a year or two, the increased size of the trunk compelled it to split into irregular pieces; and hence arises the rugged appearance of the exterior of most trees. The Plane, and some other trees which do not present this rough appearance, annually throw off the outer coat of liber and cuticle in large plates, so that their bark furnishes no criterion for discovering the age of the trees. The Currant-tree throws off its outer bark in horny rings; the Birch, in long thin ribands. The cuticle is not confined to the trunk of a tree, but invests the branches, leaves, and even the most delicate parts of the flower, being modified into hairs, down, prickles, &c., and being frequently perforated for the transmission of fluids and gases. Between the outer circle of wood and the inner layer of bark is interposed (while the tree is in a growing state) a mucilaginous fluid, called *cambium*, which, as it exhibits traces of cellular structure, is supposed to be destined to be converted into the wood and bark of the current year. It probably owes its origin to the newly-formed wood and bark with which it is in contact, but keeps up the communication with the pith by means of thin plates of cellular tissue, called *medullary rays*, or *the silver grain*, which radiate from the centre to the circumference throughout the whole length of the stem: they are very conspicuous in wainscot oak.

Exogenous plants derive their name from depositing successive layers of wood *outside* that previously formed: in these the wood nearest the centre is the oldest and hardest, and the stem is largest near the base. *Endogenous* trees, on the other hand, have no central column of pith, nor are the vascular and cellular systems defined; but every new deposit of wood is made *within* a trunk, which, when once formed, does not alter in size during any period of its existence: consequently, the stem is constantly becoming more and more compact, the most perfect and the hardest part of its substance being near the circumference. Of this class of trees we have no British specimens; but in many of the Palms, which attain a great age, the outside becomes so hard as to withstand a blow of a hatchet.

The structure of the branches of Exogenous trees is precisely similar to that of the trunk, each one consisting of a central column of pith, a medullary sheath, and a number of rings of wood and bark corresponding to the age of the branch, these various parts being severally continuous with the same organs in the trunk. Even the topmost twig is but a repetition of the main stem as it was in its infancy. The leaf, with its stalk or *petiole*, is composed of transparent, colourless cuticle, enclosing bundles of woody fibre and spiral vessels, which expand and form a network, the interstices of which are filled up with pulpy cellular tissue, usually of a green colour, and called *parenchyma*. The lower end of the petiole is articulated to the bark, and the leaf itself differs from the solid part of the trunk by being merely of temporary duration, usually falling off with a clean fracture before it begins to decay. If, however, a branch be severed from the trunk, or vitality be suddenly destroyed in any other way, the leaves generally lose the power of throwing themselves off. Hence we may frequently see, in winter, a broken branch hanging from a tree still retaining its withered foliage, when not a leaf is to be seen on the healthy branches. Leaves have been called the lungs of a plant, not from any resemblance in shape which they may be supposed to bear to that organ, but from a similarity in their functions. In the lungs of animals the blood is exposed to the action of atmospheric air, the oxygen of which is retained, and carbonic acid gas is respired, the blood itself being thus converted into proper nourishment for the animal frame: and the leaves of plants absorb through pores in their cuticle carbonic acid gas, which they in like manner decompose; carbon, the principal component of woody fibre, is deposited, and pure oxygen restored to the air. By this arrangement of the all-wise Creator of the universe, animals are incessantly breathing out a gas deleterious to themselves but essential to the growth of vegetables; and the latter, while in their growing state and exposed to the action of light, are engaged in separating from the air all that is noxious to animals, and restoring to it in its pure state that which supports their vitality. Thus, we are permitted to see, that, were not the earth occupied by the proportion of animals and vegetables which He has ordained, it would be fit for the support of neither. The leaves, too, (to borrow a simile from machinery,) serve as a safety-valve to the whole complex structure of which they form a part: if the roots absorb a greater quantity

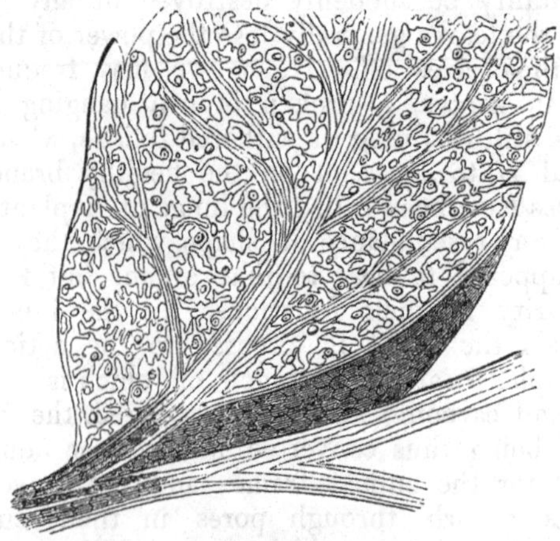

CUTICLE OF A LEAF, SHEWING THE STOMATA.

of moisture than is consistent with the well-being of the tree, the leaves are furnished with an apparatus for transmitting it in the form of vapour to the air; and if, on the other hand, the ground be parched and unable to furnish a supply, they absorb as much as will compensate for the deficiency, and return it, duly prepared for the nourishment of every organ which requires food, and so combined with other substances, and in such proportions, that whatever may be needed, be it gum, or resin, or starch, or sugar, or any other of the numberless substances which exist in vegetables, every twig, vessel, and cell is fed as it was when the earth first brought forth the fruit-tree yielding fruit *after his kind.*

The vessels nearest the upper surface of the leaf are connected with the medullary sheath: those beneath with the liber; and there is good reason to suppose that the cambium is deposited in the position which it occupies, between the bark and the wood, by the returning vessels of the leaf. The cuticle, both on the upper and under surfaces, is plentifully furnished with pores, termed *stomata,* the precise action of which is much disputed. The petiole is sometimes furnished with a leaf-like appendage termed a *stipule,* as in the Rose and many species of Willow. The angle between the base of the pedicle and the stem is called the *axil,* and is always occupied by a *bud,* which from its position is said to be *axillary.* A bud in the deciduous trees of northern climates is a rudimentary shoot enclosed within scales, which serve to protect it from cold and accidents. In young trees it usually produces a spray of leaves similar to that

in the axil of which it is situated. In many cases those buds only which are nearest to the extremity of a branch are developed: the rest appear to be reserved in case those above them should be injured by blight or other accident: but that every one is perfectly adapted for *perpetuating the individual*, may be proved, by removing from a branch in early spring all those which are evidently beginning to burst; when it will be speedily found, that the smallest and least promising bud on the branch (if it be the only one) will appropriate the nourishment intended for the rest, and burst into active life.

At a period which varies in different kinds of trees, some buds will be observed to assume a more complex structure; they no longer exclusively produce leaves of an uniform character, but make preparations for *continuing the species*. Some of the buds now contain the embryos of plants which are destined to have roots and trunks of their own. These must be nursed and protected and matured by leaves, so altered in their structure and offices, that, had we not frequent opportunities of observing them in their transition stages, we might well doubt whether they were leaves at all, and not rather distinct organs, referable to no type hitherto existing in the plant. The varieties of form to which the *flower* is subject are far too numerous to be even touched upon here. They may literally be said to be endless; for many as those are which have been described, every traveller in unexplored countries is daily adding to the list. For the sake, however, of explaining the terms which occur in these volumes, it will be necessary to mention the principal modifications of the leaf which exist in some one or other of our common flowers. After the expansion of a bud of this more complicate structure, a stem will be observed to protrude very similar to a petiole, except that it is not channelled on its upper side and is generally known by the name of *peduncle*. It is either simple or branched, and bears at its extremity, or in the latter case at the extremity of each *pedicel* (or little peduncle), several small leaves (*sepals*) united at their bases so as to form

a cup, hence called the *calyx*. The modified leaves which constitute the calyx are usually of a different form from the true leaves, but not always; the sepals of some species of Rose scarcely differ from the leaves, except in being smaller and more dilated at the base. The calyx is frequently accompanied externally by small stalkless leaves, which are termed *bracts*. They are generally more like the sepals than the true leaves, but sometimes, and especially when continued through the whole length of the peduncle, pass by an almost

CALYX LEAF OF ROSE.

imperceptible transition from one into the other. Within the calyx is the *corolla*, an assemblage of delicately modified leaves, arranged alternately with the sepals, and usually *coloured*, a word used by botanists to denote any hue but green. They are called *petals*, and, as well as the sepals, are often united throughout their lower portion into a *tube*, the expanded part being then called the *limb*. The lower part of a single petal is called the *claw*, the upper the

BRACTS, CALYX, AND COROLLA.

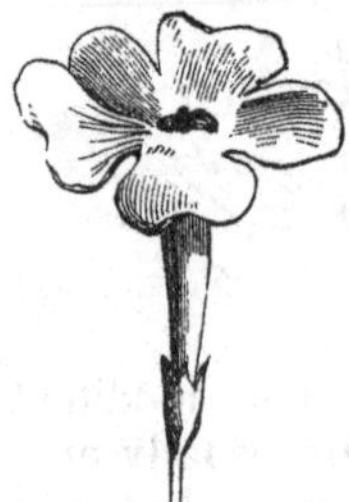

COROLLA OF ONE PETAL.

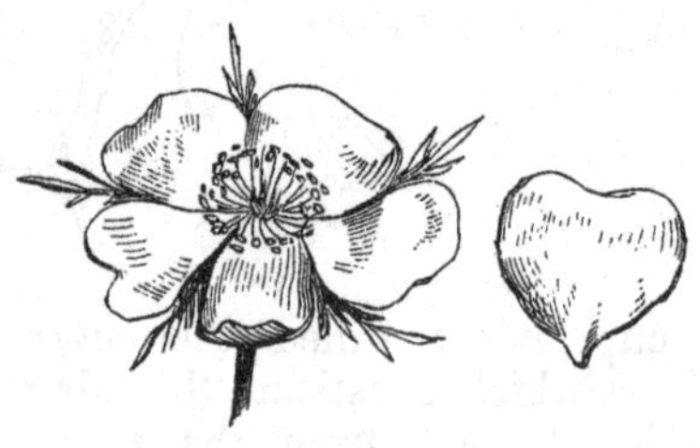

COROLLA OF FIVE PETALS. PETAL

border. Petals often bear so strong a resemblance to sepals, that it is almost impossible to distinguish one from the other, as in the tulip. Within the corolla, and sometimes cohering with it, is a row or several rows of delicate threadlike organs, called *stamens*. These are composed of three parts: the *filament*, or thread-like portion, the same in structure as the petiole of a leaf

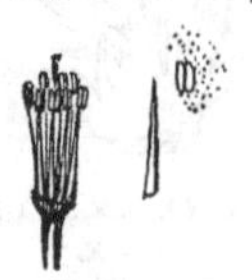

STAMENS.

but more delicate; the *anther*, a two-celled body surmounting the filament, and, when mature, usually bursting longitudinally, and allowing an escape to the *pollen*, a light dust-like substance, which is the fructifying principle of the flower. Within the stamens, and occupying a central position in the flower, is the *pistil*, which also consists of three parts, the *germen*, the *style*, and the *stigma:* the *germen*, or *ovary*, is a hollow case at the base of the pistil enclosing the seed, and finally becoming the *fruit;* the *style* is the column which rises from the ovary and supports the *stigma* or summit of the pistil. The stigma differs from all other parts of the plant, in not being covered with cuticle: it generally has a moist surface for the purpose of arresting the particles of pollen, which convey the fertilising principle through the tubes of the style to the ovary. The filament and style are not always present, and in this case the anther and stigma are said to be *sessile*, or sitting. The calyx and corolla are not essential to the perfecting of seeds; but, unless stamens and pistils are present, either in the same or different flowers, no fruit can be matured. On the number, relative lengths, combinations, and position of these essential organs, Linnæus founded his Artificial System of the arrangement of plants, the *class* being for the most part decided by reference to the stamens, the *order* being dependent on the pistils. It is now unfortunately too much the custom to decry the system of Linnæus, and to speak of his time as "the dark age of

PISTIL.

botany;" but its great inventor himself confessed it to be imperfect, and recommended it only as a substitute for some undiscovered system, which should associate plants of similar structure; his own method being open to the objection, that it brought together those which were not physiologically connected, and separated many which were closely related. Modern botanists have freely availed themselves of the discoveries of Linnæus, and have undoubtedly made considerable advances towards a Natural System, against which this objection cannot be urged, but they have neglected to tender their acknowledgments to one who did more to dissipate the gloom in which the science of Natural History was shrouded, than any, or even all, of his predecessors.

It has been already stated, that all our British trees are in their growth *exogenous;* hence they are arranged in a *class*, together with numberless herbs, shrubs, and trees, and called *Exogens*. They are distinguished by the marks given above, and by their seeds being composed of at least two lobes, called *cotyledons*, held together by a minute organ, the upper part of which, the *plumule* (a little feather), is a rudimentary stem; the lower, the *radicle*, is, as its name implies, a rudimentary root. The cotyledons are leaves, differing in shape from those afterwards developed, and serving to nourish the young plant until proper leaves are formed. When the seeds germinate, the cotyledons generally rise above the ground, bringing the plumule with them; sometimes, as in the

COTYLEDONS OF BEAN.

Windsor bean, they remain beneath the soil until their office is fulfilled, when they perish. If destroyed prematurely, the young plant dies

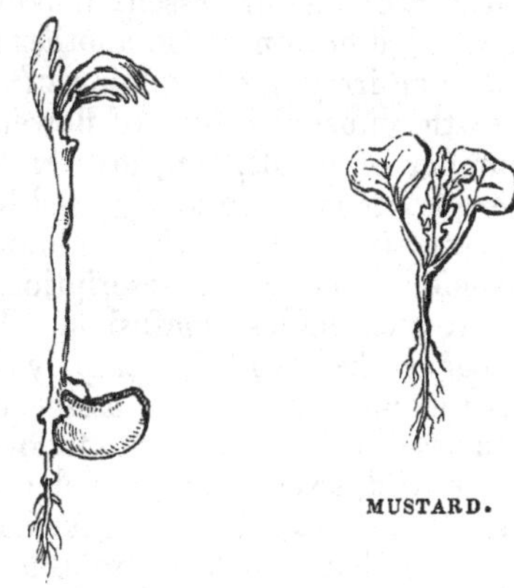

MUSTARD.

WINDSOR BEAN.

with them; in garden-mustard, for instance, they are cut as spring salad, and the plants wither away. As the number of cotyledons is nearly always two, this class is by some botanists termed *Dicotyledons.* Dicotyledons, or Exogens, are subdivided into several *sub-classes,* and these again into a multitude of *orders,* the limits of which it is not necessary to touch on in a work of this kind. Suffice it to say, that each *natural order* consists of a number of plants assembled, to a certain extent, arbitrarily, though not without regard to their similarity of structure, especially in the organs of fructification. The plants comprised in each natural order are again distributed into *genera* (families), each *genus* including all plants which resemble one another yet more closely in the essential characters of fructification. A *species* (kind) is an assemblage of individual plants, agreeing with each other in *all* essential points; and individuals which differ one from another in minor points, such as an irregular formation of leaves or mode of growth, unusual colour of flowers, extraordinary number of petals, &c., are termed *varieties.* These words are frequently used loosely in common conversation; but the habit cannot be too carefully avoided in botanical descriptions, as calculated to produce endless confusion. Throughout these pages they will be employed exclusively with the meanings above assigned, which will be rendered clearer by the following examples:—The wild, sweet-scented violet is called by botanists *Viola odorata;* the former name, *Viola,* indicating that it belongs to the *genus* so called, and being therefore termed its *generic name.* Besides the scented violet, we have in England the dog violet, the marsh violet, the pansy, and several others, all belonging to the same *genus,* and being therefore included under the name *Viola:* but the dog violet differs from the sweet-scented, in having acute sepals and leafy stems, whereas the latter has blunt sepals, and the leaves spring directly from the roots. The dog-violet is therefore a distinct *species, Viola canina.* The marsh violet and pansy differ also in important characters; they are therefore also considered distinct species, the fact being indicated by the addition of the *specific* or *trivial* names, *palustris* and *tricolor,* to the *generic* name *Viola.* The flowers of the scented violet are sometimes white and sometimes blue;

garden specimens are often tinged with pink, and still more frequently double: these characters being either unimportant or inconstant,—for blue

DOG VIOLET. SCENTED VIOLET.

flowers generally have a great tendency to sport to white, and double flowers are not perpetuated by seed,—the blue, white, pink, and double sweet violets are not considered distinct species, but mere *varieties*. Now there are many plants which bear a close resemblance to a violet in the structure of their flowers and seeds, but yet differ so far, that they cannot be reduced under the same *genus;* they are therefore placed with it in the same *natural order* called VIOLACEÆ, or VIOLETWORTS, all the genera in which differ in essential points from the genera which compose other orders, but agree with a vast number in having *two-lobed seeds* and *leaves with netted veins*, two of the characters of *Exogens*. The flower of which we have been speaking belongs, then, to the *class* EXOGENS, *natural order* VIOLACEÆ, *genus* Viola, *species* odorata, of which species it is a white, or blue, or double *variety*. In the Linnæan system, the same plant is placed in the *class* Pentandria, which comprises flowers having five stamens; and in the *order* Monogynia, which includes such of them as have but one pistil.

Every *natural* arrangement of plants professes to bring together those which most resemble each other in all respects; and the reader will see, from the following table, how far the promise is fulfilled with regard to those orders which contain British trees. The botanical characters of each order are not given, partly to avoid extending the introduction to too great a length, and partly from the difficulty which attends the execution of such a task, without employing technical terms, which would be unintelligible to the general reader.

THE LIME TREE.

TILIACÆ.

LINDENBLOOMS.

The plants belonging to this natural order are mostly trees or shrubs. They have all a mucilaginous wholesome juice, and many of them are remarkable for the toughness of the fibres of the inner bark. In a species of Aristotelia this is so strong, as to be converted into strings for musical instruments. One genus (*Corchorus*) supplies the Indians with fishing-lines and nets; and the Lime or Linden tree furnishes the material of which, in Russia, bast mats are made. Some genera produce edible berries, and the bony seeds of others are not uncommonly set in gold, and form handsome necklaces. In several instances the timber is employed for the most useful purposes. The name of the order is derived from Tilia, the Linden-tree, the only British genus.

ACERINEÆ.

MAPLES.

This is a small order, comprising only three genera, all of which are confined to the temperate regions of the globe. The more remarkable species, with their properties, will be found mentioned in the description of the Sycamore and Maple. The only British genus is Acer, the Maple, which gives name to the order.

LEAF AND FLOWERS OF THE SYCAMORE.

HORSE CHESTNUT.

HIPPOCASTANEÆ.

HORSE-CHESTNUTS.

These are trees or shrubs, furnished with irregular flowers, the petals of which never agree in number with the stamens. The few species that belong to this order grow in the north of India and North America. They are chiefly remarkable for their large seeds, which contain a considerable quantity of starch, and consequently possess nutritive properties. It is, however, asserted that the Buck-eye, or American Horse-chestnut, is a deadly poison, a bitter narcotic principle being contained in the root, leaves, and fruit. Potash also enters into the composition of the seeds, and in some instances to such an extent, that they may be used as a substitute for soap.

The Horse-chestnuts are included by some botanists in the more extensive order of SOAPWORTS, of which the following are the most remarkable examples. The Soap-tree, *Sapindus saponaria*, produces fruit which lathers freely in water, and is used in the West Indies as a substitute for soap. A pound of the fruit, it is said, will cleanse more linen than sixty pounds of soap. The distilled water of the flowers of the soapy Cupania is used by negro-women as a lotion for the face.

Other allied genera are very poisonous: a species of Paulinia, a Brazilian tree, has bark, leaves, and fruit, which abound in an acrid principle, and the blacks prepare from them an insidious poison which slowly but certainly destroys life. Several others possess intoxicating properties, and some are medicinal. From the seeds of *Paulinia sorbilis* a food called Guarana bread is prepared by the savages of Brazil, which is sold all over the country as an indispensable requisite for travellers, and a cure for many disorders. The genus *Nephelium* furnishes many of the most delicious fruits of the Indian Archipelago. The Snake-nut tree, described by Schomburgk, is a native of Demerara, and is so called from the curved embryo of the seed which, when in a state of germination, resembles a coiled snake. No plant of this order is indigenous to Britain, but several varieties of Horse-chestnut are common in parks and plantations. The Paira-tree is not unfrequently to be met with in such situations. It differs from the Horse-chestnut in having its seeds enclosed in a smooth instead of prickly case.

CELASTRINEÆ.

SPINDLE-TREES.

A rather large number of plants are included in this order, but not many of great interest. They are natives of the warmer parts of Europe, North America, and Asia, and a great number inhabit the Cape of Good Hope. A few also occur in Chili, Peru, and New Holland. Many of them possess an acrid, stimulant principle: the green leaves of one species are said to be eaten by the Arabs to produce watchfulness, and a sprig of it is believed to be a protection from the plague to the person who carries it. The only British species, the Spindle-tree, is most remarkable for its pink, lobed seed-vessels, which, in autumn, render the tree a conspicuous object. One species of Celastrus (the genus from which the order takes it name) is said to inflict very painful wounds. The English name is derived from the use made of its very compact wood.

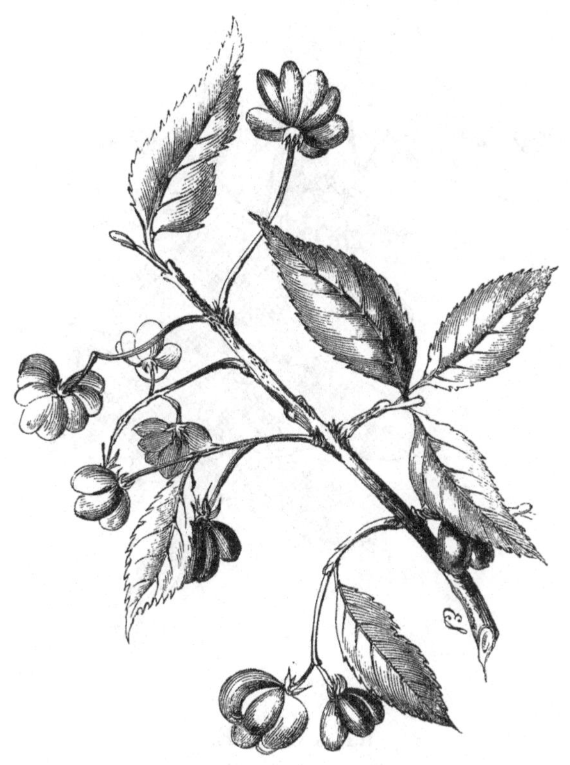

BRANCH OF THE SPINDLE-TREE.

BUCKTHORN.

RHAMNEÆ.

RHAMNADS.

This order contains between two and three hundred trees and shrubs, often thorny, which inhabit all parts of the world except the arctic zone. Some species of Ziziphus produce the jujube, well known in this country as a sweetmeat. *Ziziphus Lotus* is famous for being the plant which afforded food to the ancient Lotophagi, or Lotus-eaters.* Homer states that it was so delicious, that whatever stranger once tasted it immediately forgot his friends and native country, and desired only to dwell within its reach. It is a prickly shrub, and bears an abundance of purplish berries the size of sloes, containing large stones. The pulp is mealy and of a delicious flavour. Under the name of *seedra* or *sadr*, it still affords food to the Arabs, who separate the pulp from the stone by gently pounding the fruit in a mortar, and either convert it into a kind of bread at once, or lay it by for winter use. A kind of wine is also made from the fruit, but this will not keep more than a few days. Mungo Park, Dr. Shaw, and other travellers found the tree in abundance in many of the sandy parts of Arabia; and the latter states that the fruit, called *nabk*, is regularly exposed for sale in the markets of Barbary. *Ziziphus spina-Christi* and *Paliurus aculeatus*, prickly shrubs common in the East, are severally believed by many persons to have furnished our Blessed Saviour's crown of thorns. Only two plants of this order are indigenous to Britain, and belong to the genus Rhamnus; their berries are medicinal, but too violent in their effects to be used with safety.

* See Tennison's beautiful poem "The Lotus-eaters." The Egyptian lotus is a very different plant, being a species of water-lily, *Nymphæa Lotus.*

ROSACEÆ.

ROSEWORTS.

This extensive order, which comprises nearly all the fruit-trees of the temperate regions which are valuable to man, has been subdivided by modern botanists into Drupaceæ, Pomaceæ, Sanguisorbaceæ, and Rosaceæ. The plants of all these sub-orders are marked by bearing an indefinite number of stamens on the calyx, and agree in some other particulars.

DRUPACEÆ.

ALMONDWORTS.

Almondworts are distinguished by bearing what is technically called a drupe, that is, a fleshy or pulpy fruit enclosing a hard stone, and by the presence of prussic acid in their leaves and kernel. They are natives exclusively of the cold and temperate climates of the northern hemisphere, but in a cultivated state are diffused throughout most parts of the world. The poisonous properties of prussic acid are too well known to require any notice: yet, notwithstanding the presence of this destructive principle in the leaves and other parts of the trees belonging to this order, the fruit is with very few exceptions harmless, or even a nourishing food. The order includes the Almond, Peach, Nectarine, Apricot, all the varieties of Plum,

Cherry, and Laurel. Of these the last alone bears poisonous fruit. The trees of this order which are natural to Britain, or have been introduced, will be described as belonging to the genus Prunus, though classed by some botanists partly under Prunus and partly under Cerasus.

LEAF AND FRUIT OF THE WILD CHERRY.

FRUIT OF THE MOUNTAIN ASH.

POMACEÆ.

APPLEWORTS.

About two hundred species are included in this order, which consists entirely of trees and shrubs. They are natives of the more temperate regions of the northern hemisphere, but some of them have accompanied man in his migrations to regions most remote from their native soil. The varieties of Apple and Pear which have been produced by cultivation are beyond the power of calculation. These two are most valued for their fruit. Other British trees belonging to the order are the White-beam, Mountain Ash, Service, and Hawthorn, all more or less valuable for their beauty or the uses to which their timber may be applied. Malic acid is contained in considerable quantities in apples, and yet more abundantly in the berries of the Mountain Ash. The flowers, bark, and root of the last-named tree yield also a large quantity of prussic acid.

TAMARISCINEÆ.

TAMARISKS.

A small order, comprising bushy trees with rod-like branches and a light feathery spray, no part of the leaf being expanded into a plate, but the foliage consisting entirely of scales, which are closely pressed to the stem. Hence the trees of this order are eminently adapted for growing in exposed situations, where indeed they are usually found. The majority grow by the sea side, others are found by the banks of rivers and torrents, or in arid districts where the soil is impregnated with salt. In Arabia, one species (*Tamarix mannifera*) has its branches invested with a sweet mucilaginous substance, which is collected by the monks of Mount Sinai, and sold as a substitute for sugar.

Other species are remarkable for the large quantity of sulphate of soda which their ashes contain. The only British species is *Tamarix Gallica.*

TAMARISK.

IVY.

ARALIACEÆ.

IVYWORTS.

This order derives its name from the Aralia, or Angelica-tree, a prickly shrub introduced into our gardens from Carolina and Virginia. A species of Panax, which furnishes the famous ginseng of China, is a native of Chinese Tartary, where it has been gathered as an invaluable drug from time immemorial. In 1709 the Emperor of China gave orders to 10,000 Tartars to go in quest of the root, and to bring as much as they could find: every one was to give two pounds of the best to the Emperor, and to sell the rest for the same weight of fine silver. The roots, which are said to bear some resemblance to the human form, are gathered and dried, and enter into almost every medicine used by the Chinese and Tartars. Osbeck says that he never looked into the apothecaries' shops, but they were always selling ginseng; that both poor people and those of the highest rank made use of it, and that they boil half an ounce in their tea or soup every morning as a remedy for consumption and other diseases. Jartoux relates that the most eminent physicians of China have written volumes on the medicinal powers of this plant, asserting that it gives immediate relief in almost every kind of disorder. European physicians, however, seem to doubt its efficacy, at least in this climate. A remarkable plant belonging to this order is *Gunneia scabra,* found by Darwin, growing on the sandstone cliffs of Chiloe. He describes it as somewhat resembling rhubarb on a gigantic scale, each plant producing four or five leaves nearly eight feet in diameter. Of the genus Hedera, (Ivy,) fifty-two species are enumerated by Don, but many of these are referred by other authors to Aralia, &c. The Ivies are either climbing shrubs, like our own familiar plant, or grow to the height of fifty feet without support. The only British plant besides Ivy belonging to this order is the Moschatell, (*Adoxa Moschatellina,*) a humble plant which appears early in spring, on shady moist banks, and may be distinguished by its delicate foliage, and globular heads of pale green flowers, the whole plant having a strong smell of musk, whence it derives its name.

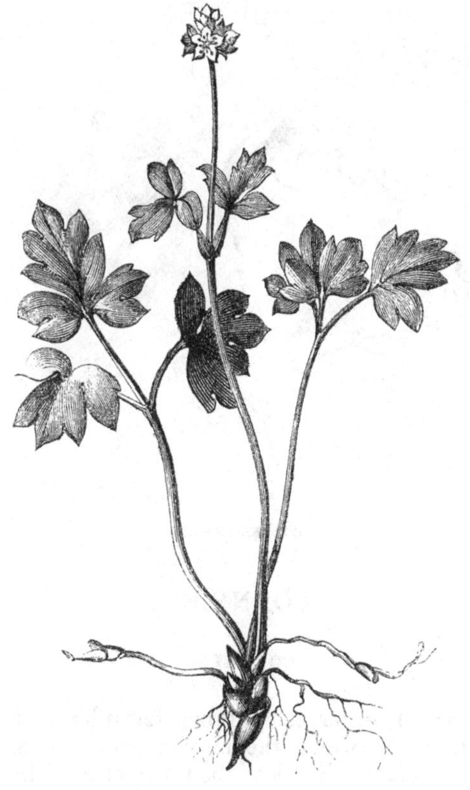

THE MOSCHATELL.

CORNEL-TREE.

CORNEÆ.

CORNELS.

A small order, containing few plants of interest, among which the most remarkable is the Cornel-tree, to be described hereafter. In the United States several species are found, the bark of which is a powerful tonic, ranking in utility next to Peruvian bark. *Benthamia fragifera*, a handsome shrub from the mountains of Nepal, was introduced into England in 1825. In Cornwall, where it was first raised from seed, it flowers and bears fruit freely, and forms a pleasing addition to the shrubbery. Two species of Cornus are indigenous to Britain: one, *C. sanguinea*, a shrub, distinguished by its blood-red twigs; the other, *C. Suecica*, a herbaceous plant growing in the mountainous parts of England and Scotland, the berries of which are said to increase the appetite, whence its Highland name, *Lus-a-chrasis*, or, plant of gluttony. The Cornus of the ancients was the present Cornelian cherry, *Cornus mascula*, whose little clusters of yellow starry flowers are among the earliest heralds of spring. Its fruit is like a small plum, with a very austere flesh, but after keeping becomes sub-acid. The Turks still use it in the manufacture of sherbet. A similar species is commonly cultivated in Japan for the sake of its fruit, which is a constant ingredient in the fever drinks of the country.

CAPRIFOLIACEÆ.

CAPRIFOILS.

In this order are associated a number of plants very unequal in size, and perhaps too dissimilar in structure, for here, with the Elder, Woodbine, Guelder Rose, and Wayfaring Tree, we find the elegant Linnæa, "the little northern plant, long overlooked, depressed, abject, flowering early," which Linnæus himself selected as therefore most appropriate to transmit his name to posterity. Most of the plants of this order are confined to the temperate regions of

LINNÆA.

the northern hemisphere. The roasted berries of *Triosteum perfoliatum* have been used as a substitute for coffee. *Leycesteria formosa*, a beautiful shrub from the mountains of Nepal, where it flowers at an elevation of from 6000 to 8000 feet, is now becoming a common ornament in our gardens. It is most attractive when

HONEYSUCKLE OR WOODBINE.

in a flowering state, from the contrast of the deep green hue of its stem and leaves with the purple colour of its floral leaves and berries. The name of the order is from *Lonicera caprifolium*, a species of Honeysuckle or Woodbine.

ILICINEÆ.

Hollyworts.

The only European plant belonging to this order is the Holly; but in North and South America, the West Indies, and South Africa, other species occur, all of which agree in being evergreens, and in presenting a structure of flower and fruit more or less resembling that of our Holly. The most famous of all is the Paraguay Holly, which furnishes the *maté* so extensively used in South America, an account of which will be given hereafter. The bark and berries of some species are highly astringent, and valuable as medicines. An infusion of *Ilex vomitoria* is used by some of the southern tribes of American Indians at the opening of their councils, to produce the effect indicated by the name. After having indulged in this strange ceremony for two or three successive days, the senators proceed to their deliberations. Several other species are used as tea.

OLIVE BRANCH.

OLEACEÆ.

Oliveworts.

In this order are assembled a variety of trees and shrubs, which the mere casual observer would perhaps think very ill assorted, as the Olive, Ash, and Lilac. Twenty-six species of Olive are enumerated by Don, ("Gardener's Dictionary,") of which only one, *Olea sativa*, is cultivated for the oil which it affords. It is said to have come originally from Asia, and grows abundantly about Aleppo and Lebanon. It is also naturalised in the south of France, in Italy, and in Spain. The Olive is remarkable for the very great age which it attains, some plantations in Italy being supposed to have been in existence in the time of Pliny, eighteen centuries since. It is an evergreen tree, but seldom exceeds thirty feet in height. Its productiveness increases with its age: a modern authoress mentions a tree near Gerecomio, which yielded 240 quarts of oil in one season; yet its trunk was quite hollow, and its empty shell seemed to have barely enough hold in the ground to secure it against mountain storms. The olive flourishes best on limestone soils, and in the vicinity of the sea. The fruit is an oblong stone enclosed in a fleshy pulp, ripening in November, when it is of a reddish-purple colour. As soon as gathered, it is placed in a mill, so contrived as to separate the fleshy part without breaking the stones. The pulp thus prepared is put into bags made of rushes, and submitted to a gentle pressure; the liquid which flows off is pure oil of the best quality. The residue is then beaten up and moistened with water, and returned to the press; upon which there flows out oil of an inferior quality, mixed with water. What remains after this process is again mixed with water, and set to ferment, after which it is again pressed, and a coarser oil is produced, valuable principally to the soap-boiler. With us oil is not much used as food; but in Italy and Spain it takes the place of butter and cream. Captain Head relates that an English company was formed to supply the Spanish population of Buenos Ayres with butter; but when everything seemed to favour the scheme, the speculators were thwarted by the discovery that oil was infinitely preferred. Numerous passages in the Scriptures, in which mention of it occurs in connexion with corn and wine, prove that it was a staple article of food in the East from the remotest antiquity. Then, too, as now, it was extensively used in medicine and surgery, and among all civilised nations of the eastern continent it has been regarded as the emblem of peace. So highly did the Greeks value it, that they ascribed its production to the tutelary goddess of Athens, and pointed to the identical tree which they pretended to have sprung from the ground at her bidding; and Pliny, the Roman naturalist, pronounces it to be of greater value than the vine. Pickled olives are prepared from unripe fruit, by soaking them in water, and then bottling them in salt and water. It is singular, that Evelyn, who could not have been aware of the fact, that the Olive and the Ash were kindred genera, recommends the seed-vessels of the latter tree to be treated similarly, for, "being pickled tender," he says, "they afford a delicate salading." The custom of grafting the Olive which, according to Pliny, was as well known to the Romans in his day as it was when St. Paul addressed his epistle to the same people, is now rarely if ever practised, it having been discovered that the tree may readily be propagated by cuttings or slices of the root. Many varieties of the Olive have been produced by cultivation, as in the Apple, Pear, and other valuable fruit-trees. The well-known ornamental shrub Lilac is a native of Persia, whence it was imported in the sixteenth century; it is so called from *lilac* or *lilag*, the Persian for a flower. Though now so plentiful as to have given name to a colour, in the year 1597 it was a great rarity. It possesses medicinal virtues, and it is stated on good authority, that, in a part of the province of Berri, which is marshy and exceedingly unwholesome, the peasants employ no other remedy for the intermittent fever which prevails there. Of the Ash, Don enumerates thirty-seven species, of which one only is a native of Britain. The allied genus *Ornus*, or Flowering Ash, contains several species which afford the substance called *manna*, of which more hereafter. Privet, a common hedge shrub, also belongs to this order. Dissimilar as many of these plants may appear, it is remarkable that they will all graft on one another, a fact which demonstrates the analogy of their juices and fibres. Thus, not only will the Olive graft on the Wild Olive, but on the Ash; and the Lilac will take on the same tree; so that it is not beyond the skill of the gardener to produce a tree with the trunk of an Ash, bearing from the same root branches of Olive, Lilac, Privet, and Phyllirea, &c.

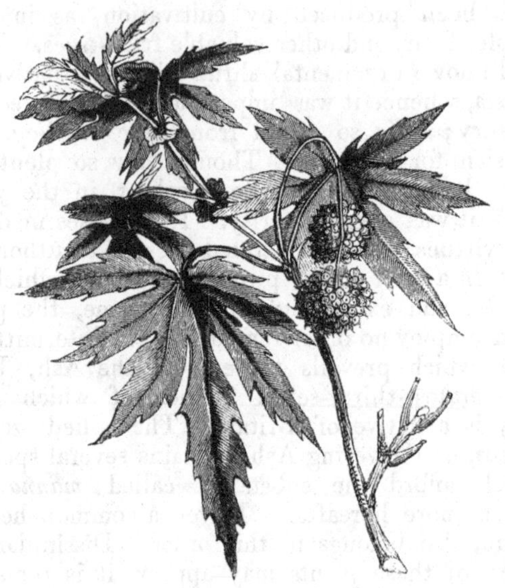

LEAF AND BLOSSOM OF ORIENTAL PLANE.

PLATANACEÆ.

Planes.

This order is known in England by two noble timber trees, which are natives respectively of the eastern parts of Europe and of North America. The Oriental Plane has ever been a favourite tree in the countries which it inhabits; and the Occidental Plane is highly prized even in its native land of forests. Both species are frequently to be met with in English parks and plantations; but the other two or three species which constitute the genus and order require no further notice.

EUPHORBIACEÆ.

Spurgeworts.

A large order, containing nearly 200 genera and 2500 species, distributed over most of the tropical and temperate regions of the globe, especially the warmer parts of America. They are either trees, shrubs, or herbs, and some kinds have the external habit of the cactus tribe. Among so numerous an assemblage of plants, we should expect to find a great dissimilarity of properties, which, indeed, exists to a certain extent; yet nearly all agree in being furnished with a juice, often milky, which is highly acrid, narcotic, or corrosive, the intensity of the poisonous property being usually proportionate to the abundance of the juice. Of the genus *Euphorbia*, Spurge, which gives name to the order, ten or twelve species are natives of Britain. The British Spurges are all herbaceous, and remarkable for the singular structure of their green flowers, and their acrid milky juice, which exudes plentifully when either the stems or leaves are wounded. A small quantity of this placed upon the tongue produces a burning heat in the mouth and throat, which continues for many hours. The unpleasant sensation may be allayed by frequent draughts of milk. The roots of several of the common kinds enter into the composition of some of the quack fever medicines; but they are too violent in their action to be used with safety. The Irish Spurge is extensively used by the peasants of Kerry for poisoning, or rather stupifying, fish. So powerful are its effects, that a small creel or basket filled with the bruised plant suffices to poison the fish for several miles down a river. *Euphorbia Lathyris* is sometimes, though erroneously, called in England the caper-plant. Its unripe seeds are pickled, and form a dangerous substitute for genuine capers, which are the unexpanded flower-buds of *Caparis spinosa*, a shrub indigenous to the most southern countries of Europe. Among the foreign Spurges, some species furnish both the African and American savages with a deadly poison for their arrows. Another, called

SPRIG OF BOX.

in India *Tirucalli*, furnishes an acrid juice, which is used in its fresh state for raising blisters. Other

AMYGDALOIDES EUPHORBIA.

kinds are used in various parts of the world as medicines, but require to be administered with caution. The gum resin, Euphorbium, of chemists, is procured from three species growing in Africa and the Canaries, by wounding the stems, and collecting in leathern bags the sap which exudes. It is an acrid poison, highly inflammable, and so violent in its effects, as to produce severe inflammation of the nostrils if those who are employed in powdering it do not guard themselves from its dust. Pliny relates that the plant was discovered by King Juba, and named by him after his physician, Euphorbus. The Manchineel tree (*Hippomane Mancinella*) is said to be so poisonous, that persons have died from merely sleeping beneath its shade. Its juice is pure white, and a single drop of it falling upon the skin burns like fire, forming an ulcer often difficult to heal. The fruit, which is beautiful and looks like an apple, contains a similar fluid, but in a milder form; the burning it causes in the lips of those who bite it guards the careless from the danger of eating it. *Jatropha Manihot*, or Manioc, is a shrub about six feet high, indigenous to the West Indies and South America, abounding in a milky juice of so poisonous a nature, that it has been known to occasion death in a few minutes. The poisonous principle, however, may be dissipated by heat, after which process the root may be converted into the most nourishing food. It is grated into a pulp, and subjected to a heavy pressure until all the juice is expressed. The residue, called *cassava*, requires no further preparation, being simply baked in the form of thin cakes on a hot iron hearth. This bread is so palatable to those who are accustomed to it, as to be preferred to that made from wheaten flour; and Creole families, who have changed their residence to Europe, frequently supply themselves with it at some trouble and expense. The fresh juice is highly poisonous; but, if boiled with meat and

JATROPHA MANIHOT.

seasoned, it makes an excellent soup which is wholesome and nutritious. The heat of the sun even is sufficient to dissipate the noxious properties, for, if it be sliced and exposed for some hours to the direct rays of the sun, cattle may eat it with perfect safety. The roots are sometimes eaten by the Indians, simply roasted, without being previously submitted to the process of grating and expressing the juice. They also use the juice for poisoning their arrows, and were acquainted with the art of converting it into an intoxicating liquid before they were visited by Europeans. By washing the pulp in water and suffering the latter to stand, a sediment of starch is produced, which, under the name of *tapioca,* is extensively imported into Europe, where it is used for all the purposes to which arrow-root and sago are applied. It is light, digestible, and nourishing, so much so, indeed, that half a pound a day is said to be sufficient to support a healthy man. Caoutchouc, or India-rubber, is a well-known elastic gum, furnished in greater or less abundance by many plants of this order, but especially by a South American tree, *Siphonia* or *Hevea elastica.* In order to obtain it, the trees are punctured in the rainy season, upon which a thick juice of a yellowish colour exudes, which becomes darker by exposure to the air. If kept in well closed bottles, it will retain its fluidity for a considerable time; but, if heated or exposed to the air in thin films, its moisture evaporates, and it quickly assumes the form under which we are acquainted with it. By the Indians of South America it is applied in successive layers to models of clay, and dried in the smoke of a fire, until it has acquired the requisite thickness; the clay is then crushed and shaken out. They make boots of it which water cannot penetrate, and which, when smoked, have the appearance of real leather. Bottles are also made of it, to the necks of which are fastened hollow reeds, so that the liquor contained in them may be violently forced out by pressure. One of these filled with water is always presented to each of the guests at their entertainments, who never fails to make use of it by forcing the water into his mouth before eating. Hence the tree which produces the gum is sometimes called the Syringe-tree. Caoutchouc is soluble in ether, and in the essential oils of turpentine and lavender; and even if boiled for a quarter of an hour in water, it may be so far dissolved, that the edges will firmly unite. If a bottle be soaked in sulphuric ether until quite soft, it may be inflated until it becomes so thin as to be transparent, and sufficiently light to ascend when filled with hydrogen gas. Waterproof cloth is manufactured in Europe by dissolving it in the oil distilled from gas-tar, and spreading the mixture on the surface of a piece of cloth, upon which a similar piece is then extended, and the whole passed between a pair of rollers. The largest India-rubber trees grow on the banks of the river Amazon, where they attain a great height, being at the same time perfectly straight, and bearing no branches except at the top, which is but small, covering no more than a circumference of ten feet. The leaves are large, tough, and leathery, green above, and whitish beneath. The seeds contain a thick oil, which answers the purpose of butter in the cookery of the country. The fragrant aromatic bark called cascarilla is produced by a shrub belonging to this order, *Croton Eleutheria*, a native of the Bahamas, and by other species of *Croton* indigenous to the West Indies and South America. Croton oil is the product of *Croton Tiglium*, and is so violent a medicine, as to be rarely administered until all other remedies have failed. Castor-oil is expressed from the seeds of *Ricinus communis*, an African tree, frequently to be met with in English gardens under the name of *Palma Christi*, where, however, it only attains the rank of an annual herbaceous plant. The Box is the only British tree belonging to this order, of the poisonous properties of which it partakes, though to a limited extent. In some parts of Persia it is very abundant; and in these districts it is found impossible to keep camels, as the animals are very fond of browsing on the leaves, which kill them. The Dog-Mercury (*Mercurialis perennis*) is an herbaceous plant, common in our woods, and an active poison; another species, *M. annua*, is less frequently met with, and, though poisonous, is not so virulent as the other species.

LEAF AND BLOSSOM OF THE ELM.

ULMACEÆ.

Elmworts.

An order composed of trees and shrubs which inhabit the temperate parts of Asia, North America, and Europe, and are often valuable timber trees. The Elm is the only British genus, which will be described hereafter.

e

LEAF AND BLOSSOM OF THE WALNUT.

JUGLANDACEÆ.

JUGLANDS.

THE Walnut is the only tree belonging to this order with which we are familiar in England. It is a native of Persia and Cashmere, whence it was probably introduced before the beginning of the sixteenth century. Of the Hickory tree there are many species, all natives of North America, where they are greatly valued for their nuts, as well as for their timber, which is of great weight, strength, and tenacity. It is used for various purposes, especially for cask-hoops, and the rings by which the sails of vessels are attached to the mast. When burnt, it consumes slowly, and gives out a great heat; hence it is considered the best wood for fuel. The timber of the various kinds of Hickory, when stripped of the bark, is said to be so much alike, as not to be distinguished by the most practised eye. The structure of the nut closely resembles that of the Walnut, but is not deeply furrowed like the fruit of the latter tree. *Engelhardtia spicata*, a Java tree which attains a height of 200 feet, is used for cart wheels, which are cut out of a single horizontal block. Most of the trees of this order abound in a watery resinous juice.

AMENTACEÆ.

Amentals.

This order derives its name from *amentum*, a catkin, the term used by botanists to designate the form of inflorescence when a number of flowers destitute of calyx and corolla are arranged along a common stalk, which falls off in one piece, either after flowering or after the ripening of the fruit. The plants of which this important order is composed agree in bearing their flowers in catkins, but nevertheless differ so materially in other respects, that they have been subdivided into distinct sub-orders or tribes. Of these some contain no British genera: those in which indigenous examples occur, are—

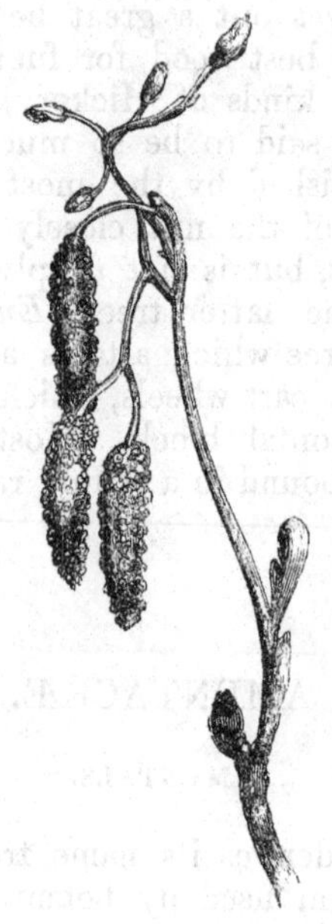

BLOSSOM OF THE ALDER.

BETULACEÆ,

BIRCHWORTS,

containing only *Betula*, the Birch; and *Alnus*, the Alder:

BLOSSOM OF THE CRACK WILLOW.

SALICACEÆ,

WILLOWWORTS,

comprising *Salix*, the Willow; and *Populus*, the Poplar.

CHESTNUT.

CUPULIFERÆ, *or* CORYLACEÆ,

MASTWORTS.

This last division is the largest and most important, and is well-known in Britain by *Carpinus*, the Hornbeam; *Corylus*, the Hazel; *Fagus*, the Beech; *Castanea*, the Chestnut; and *Quercus*, the Oak. The most remarkable species belonging to these several genera will be noticed under their respective heads.

BEECH-MAST.

SPRIG OF SCOTCH FIR.

CONIFERÆ.

Conifers.

This order derives its name from the peculiar kind of seed-vessel, called a "cone," produced by the Fir and other allied genera. Conifers are distinguished by this character, by their needle-like leaves, by the vessels of their wood being perforated with numerous dots, and the apparently imperfect structure of the pistil in the perfect flower. "No order can be named of more universal importance to mankind than this, whether we view it with reference to its timber or its secretions. Gigantic in size, rapid in growth, noble in aspect, robust in constitution, these trees form a considerable portion of woods or plantations in cultivated countries, and of forests where nature remains in temperate countries in a savage state. Their timber, in commerce, is known under the names of deal, fir, pine, and cedar, and is principally the wood of the Spruce, the Larch, the Scotch Fir, the Weymouth Pine, and the Virginian Cedar; but others are of at least equal, if not greater value. *Pinus palustris* is the Virginian Pine, so largely employed in the navy for masts. The gates of Constantinople, famous for having stood from the time of Constantine to that of Pope Eugene IV., a period of eleven hundred years, were of Cypress. The Norfolk Island Pine (*Araucaria excelsa*) and the Kaurie tree of New Zealand attain the height of 200 feet, and the latter yields an invaluable, light, compact wood, free from knots, from which the finest masts in the navy are now prepared. But they are both surpassed by the stupendous pines of north-west America, one of which, *Pinus Lambertiana*, is reported to attain the height of 230 feet; and another, *Abies Douglasii*, to equal or even to exceed it." (*Lindley.*) Great though their value be as timber, they are yet more valuable for their copious secretion of substances useful in the arts and sciences. Pitch, tar, turpentine, resin, Burgundy pitch, Hungarian balsam, Canada balsam, &c., are furnished by some one or more of these trees; and the seeds of the larger kinds are edible and nourishing. The Scotch Fir is a British example of this order. The Juniper, the berry-like cones of which are used for flavouring hollands, extends over the greater part of Europe and North America, and is likewise found in some of the mountainous parts of Asia.

TAXACEÆ.

Taxads.

This order, which derives its name from *Taxus*, the Yew, is by some botanists united with the preceding, which it resembles in most respects, except the structure of its fruit, and its repeatedly divided branches. Occasionally, also, the leaves expand, and form veins, as in the *Salisburia*. The only species belonging to this order found in Europe is the Common Yew. Other species occur in the milder climates of a great part of the world, and within the Tropics seek a congenial climate high up in the mountains. Some of them are valuable as timber trees; and one, *Dacridium taxifolium*, a New Zealand tree attains a height of 200 feet. They are resinous, like the Conifers, and, with the exception of the Common Yew, are harmless.

THE

FOREST TREES OF BRITAIN.

THE OAK.

QUERCUS ROBUR—QUERCUS SESSILIFLORA.

Natural order—AMENTACEÆ.

Class—MONŒCIA. *Order*—POLYANDRIA.

No one who considers the Lion to be king of beasts, and the Eagle king of birds, will think of assigning the sovereignty of British Trees to any other than the Oak. Within the tropics, where Nature performs all her works on a scale of magnificence unrivalled elsewhere, the stately Palm, uplifting its leafy canopy on a shaft two hundred feet in height; the Banyan, forming with its countless trunks a forest in itself; the Baobab, a tree venerable four thousand years ago: each of these may assert its claim to the kingly title. But, in England, the country of green fields, in which men labour among their oxen and their sheep; of lordly parks, with their broad smooth lawns and clustering trees; of narrow church-paths winding along by the side of brilliant streamlets, across flowery meadows, and through woods offering a shade from the heat, and a shelter from the storm, here the Oak reigns paramount. In truth he is a kingly tree, the emblem of majesty, strength, and durability. To what remote ages are we carried back—to what varying scenes are we introduced when we search for the first appearance of this patriarch in the pages of history! Under the Oaks of Mamre,* according to Jewish traditions, the father of the faithful reared his tabernacle, and meditated on another, that is a heavenly, country which God had prepared for him. One of these very trees was long looked upon with veneration by the Israelites, and (according to St. Jerome) was in existence in the reign of the Emperor Constantine two thousand years afterwards.†

Near Shechem there stood also a tree of the same species, which probably was remarkable for its size, being called in Genesis xxxv. 4, "The Oak which was by Shechem." Thus early, too, does it appear to have been marked with some peculiar sacredness, for it was chosen as a meet

* It should be borne in mind that the Oak of the Holy Scriptures is not identical with the British Oak, but, as will be seen hereafter, is either the Evergreen Oak (*Quercus Ilex*), or nearly resembles it. Celsius and other writers after him are of opinion that the tree alluded to is the Terebinth, or Turpentine-tree. It is difficult, however, for the reader of the English version of the Bible to connect the name with any other notion than that of a tree agreeing closely in character with the Oak of his own country. Whatever may be the botanical difference between the two, it is still "*the* Oak" of Palestine as much as Quercus Robur is "*the* Oak" of Britain.

† Mamre is remarkable in Sacred History for Abraham's entertaining there three angels under an Oak, which Oak also became very famous in after ages; insomuch that superstitious worship was performed there. This the great Constantine, esteemed the first Christian emperor of Rome, put a stop to by a letter written to Eusebius, bishop of Cesarea, in Palestine, for that purpose.—*Heming's Scripture Geography.*

shelter for the grave of Deborah, Rebekah's nurse (verse 8th); the particular tree being afterwards distinguished by a set name, "Allon-bachuth," or, the Oak of Weeping.*

It is here worthy of notice that in Genesis xii. 6, the passage which is in our version rendered "The plain of Moreh" is in the Septuagint rendered "The high Oak."† It is not, therefore, improbable that this Oak, or grove of Oaks, was first consecrated to God by the priestly worship of Abraham, and retained its sacred character until at least the time of Abimelech.‡ It must not be objected that the period is too long (nearly six hundred years) to assign as the duration of one tree; for, as we shall see hereafter, there is evidence of Oak trees actually existing which have attained nearly double that age.

In our own country we well know that any building or tree connected with the history of a person of note, who lived in remote ages, is regarded with universal interest. Who, for example, has not heard and thought of Shakspeare's Mulberry-tree, and Charles the Second's Oak? Probably, then, the Israelites, on their restoration to the land of Canaan, regarded with much the same feelings "the Oak which was by Shechem,"

* The difficulty of identifying the plants mentioned in the sacred volume appears to be increased in the present instance by the similarity of the names *elah* and *allon*. In Genesis xxxv. both words occur, and are rendered in our version "the Oak." In Isaiah vi. 13, they occur in juxtaposition: in this passage Coverdale translates *elah* "the Terebinth," *allon* "the Oak:" the authorized version giving *elah* "the Teil Tree," *allon* "the Oak." Canon Rogers is of opinion that *allon* should always be thus rendered.

† Τὴν δρῦν τὴν ὑψηλήν.

‡ Judges ix. 6, marginal reading.

as connecting their own history with what God had done for their forefathers before the captivity in Egypt; more particularly as the Patriarchs, in token of their faith, had not erected for themselves permanent habitations, but dwelt in tents, of which no vestige could well remain. How probable is it that the pious Israelites resorted to this tree to talk over among themselves, and to repeat to their children, the incidents of their perilous wanderings in the wilderness, and the wonders which the Lord had wrought for them!

And what tree could Joshua have had greater reason to choose than this, when he "gathered all the tribes of Israel at Shechem," and "set up there under an Oak" a stone intended to commemorate the solemn renewal of their allegiance to God? These pious motives did not, however, long continue in operation. Scarcely were the elders dead who had "outlived Joshua, and had known all the works of the Lord that He had done for Israel," when the groves were resorted to for the worship of false gods: "under every green tree, and under every thick Oak, they did offer sweet savour to all their idols;" they "burnt incense upon the hills under Oaks," choosing the wood of "the Cypress and Oak to make a god."

"It is natural," says Evelyn, "for man to feel an awful and religious terror when placed in the centre of a thick wood; on which account, in all ages, such places have been chosen for the celebration of religious ceremonies." But, to trace by what degrees this pious feeling degenerated into dangerous superstition belongs rather to the historian of men than of trees: I will not, therefore, pursue the subject any further.

In European countries the Oak was an important tree at a very early age, being valued for its fruit. In Asia the estimation in which it was held appears to have had some other origin, for, although we read in the Sacred Volume of "dates, almonds," &c., being used as articles of food, no such mention is made of acorns; nor is it probable that they were ever eaten by men in a country naturally affording fruits so much more palatable. But in Greece and Italy, before agriculture was invented or introduced, acorns held an important place among the more savoury viands of the inhabitants. The traditions of the poets tell us that strawberries, blackberries, corneils, and acorns, were the homely fare of the first inhabitants of these countries; of which, acorns must have been the most valuable, as, being of a less perishable nature than the rest, they would bear being stored away for winter use. For this reason, perhaps, it was that the Greeks believed that, of all the trees with which they were acquainted, the Oak was the first created. We need not, then, wonder that, holding this belief in its antiquity and extreme usefulness, they regarded it with veneration, and, in their ignorance of Divine Revelation, entertained the, to us extravagant, notion that the Deity chose it as a medium for making known his will to man. At the same time, it is much to be wondered at that the Israelites, who had not the excuse of ignorance, should have fallen into nearly the same fatal error, and that too with respect to the very same tree. The Oak grove at Dodona in Epirus was long resorted to by the inhabitants of the whole of Greece when they wished to inquire the will of their imaginary god, Jupiter; and we have seen that the Israelites resorted to the Oak woods of Palestine with a similar object.

Let us take warning from their example, and be careful that, with whatsoever reverence we approach the works of Nature, we forget not that they are the works of the God of Nature; and that they were planted by Him that "we may see and know, and consider, and understand together that the hand of the Lord hath done this, and the Holy One of Israel hath created it."

Baal, the false god of the Canaanites, is considered by learned men to be identical with the Roman Saturn, the Celtic Yiaoul, and the British Yule, whose festival was kept at the time when we celebrate Christmas. (You see how we are entangled in this melancholy maze of errors.) By one of these nations this *name* was worshipped as significant of the god of fire; by another it was identified with the sun; by another venerated under the form of an Oak. Its priests, who were called "Druids," professed to maintain perpetual fire, and once every year all the fires belonging to the people were extinguished, and relighted from the sacred fire of the Druids. This was the origin of the Yule-log, with which, even so lately as the commencement of the present century, the Christmas fire, in some parts of the country, was always kindled, and is even now in Devonshire and Yorkshire; a fresh log being thrown on and lighted, but taken off before it was consumed, and reserved to kindle the Christmas fire of the following year. The Yule-log was generally of Oak, though sometimes of Ash; and as the ancient Britons believed that it was essential for their hearth-fires to be renewed every year from the sacred fire of the Druids, so their descendants thought that some misfortune would befal them if any accident happened to the Yule-log. The worship of the Druids was generally performed under an Oak; and a heap of stones was erected, on which the sacred fire was kindled, which was called a "cairn," as Professor Burnet says, from *kern*, an acorn.*

The misletoe was held in great reverence, and, as it was not commonly found on the Oak, solemn ceremonies attended the search for it. When all was prepared, (the misletoe having been, no doubt, previously found by some of the assistants,) the Druids went forth, clad in white robes, to search for the sacred plant, and when it was discovered one of the Druids ascended the tree, and gathered it with great ceremony, separating it from the Oak with a golden knife. The misletoe was always cut at a particular age of the moon, at the beginning of the year, and it was only sought for when the Druids pretended to have had visions directing them to seek it. When a great length of time elapsed without this happening, or if the misletoe chanced to fall to the ground, it was considered as an omen that some great misfortune would befal the nation.

The well-known chorus of "Hey derry down," according to Professor Burnet, was a Druidic chant, signifying literally, "In a circle the Oak

* This etymology, however, is doubtful, and must be received with caution. *Caini* usually signifies "a rock;" the Hebrew *keren* has usually the same sense.

move around." Criminals were tried under an Oak tree, the judges being seated under the tree, and the culprit placed in a circle made by the chief Druid's wand. The Saxons also held their national meetings under an Oak; and the celebrated conference between the Saxons and the Britons, after the invasion of the former, was held under the Oaks of Dartmoor. The wood of the Oak was appropriated to the most memorable uses. King Arthur's round table was made of it, as was the cradle of Edward II. when he was born at Caernarvon Castle; this sacred wood being chosen in the hope of conciliating the feelings of the Welch, who still retained the prejudices of their ancestors, the ancient Britons. It was considered unlucky to cut down any celebrated tree; and Evelyn gravely relates a story of two men, who cut down the Vicar's Oak, in Surrey; one losing his eye, and the other breaking his leg, soon after.

The Oaks of Dartmoor, in Devonshire, mentioned in the above extract, have now nearly disappeared. In one spot only is there any vestige of what was once doubtless a favourite gathering-place of the Druids. This spot, called Wistman's Wood, is situated on Dartmoor, about a mile above Two-Bridges, on the left bank of the river. Imagine a mountain stream creeping slowly among blocks of moss-stained granite; on either side extends a piece of flat boggy ground to an inconsiderable distance; and at the extremity of these the hills rise to the height of two or three hundred feet, capped here and there in the distance with *tors*, or rugged summits of granite. The hill side is confusedly heaped with blocks of the same stone, and it is in the interstices between

OAK IN WISTMAN'S WOOD.

these that the trees composing Wistman's Wood have chosen to fix their habitations—a colony of patriarchs in a wilderness. The wood itself forms a ragged and interrupted belt, of about half a mile in length, including some straggling trees, separated at long intervals. The best way of approaching it is from above, for by so doing one may without difficulty obtain a pretty good view of the whole at once, and plunge in among the trees at pleasure. The trees are all Oaks, from ten to fourteen feet high, gnarled, knotted, and twisted even beyond the usual characteristic of that tree. The trunks vary from two to five feet in circumference. One which was measured consisted of three trunks, branched just above the base, each bole being about three feet in circumference. But by far the strangest peculiarity is, that all the branches, with the exception (and this not always) of the extreme spires, are matted with deep beds of moss, principally Anomodon curtipendulum, in fine fructification. Some idea of the denseness of this extraordinary integument may be formed from the fact that the moss is, in most cases, from ten to twelve inches in thickness, when the diameter of the branch does not exceed an inch and a half. It seems very probable that the superincumbent weight may operate in producing the depressed character of growth: certain it is, that a single holly-tree, near the centre of the wood, which is free from parasites, has attained the height of twenty feet, and towers above his pigmy companions, like some tall pine in a wood of ordinary growth. When first we saw this tree, indeed, having nothing to compare it with of definite size and shape but the surrounding Oaks, we fancied that it was a fir-tree, and the Oaks borrowed from it, by comparison, a dignity not their own. On a rough guess, there are from 300 to 500 veteran trees in the wood, and, as we were glad to find, a great number of saplings.*

How heavily
That old wood sleeps in the sunshine—not a leaf
Is twinkling—not a wing is seen to move
Within it; but below, a mountain stream
Conflicting with the rocks, is ever heard
Cheering the drowsy noon. Thy guardian Oaks,
My country, are thy boast—a giant race,
And undegenerate still; but of this grove,
This pigmy grove, not one has climbed the air
So emulously that its loftiest branch
May brush the traveller's brow. The twisted roots
Have clasped in search of nourishment the rocks,
And straggled wide, and pierced the stony soil
In vain; denied maternal succour, here
A dwarfish race has risen. Round the boughs,
Hoary and feeble, and around the trunks,
With grasp destructive, feeding on the life
That lingers yet, the ivy winds, and moss
Of growth enormous. E'en the dull wild weed
Has fix'd itself upon the very crown
Of many an ancient Oak; and thus, refused
By Nature kindly aid, dishonoured, old,
Dreary in aspect—silently decays
The lonely wood of Wistman.

CARRINGTON'S *Dartmoor*, p. 56.

The description of the Oaks in Wistman's Wood given above does not, however, at all accord with the usual character of the tree, which certainly is, under favourable circumstances, an apt emblem of stateliness, majesty, strength, and durability. Evelyn, after enumerating several remarkable trees, records of which have been transmitted to

* This account of Wistman's Wood was written for a Botanical Journal, "The Phytologist," January 11th, 1845.

us, says, "What goodly trees were of old adored and consecrated by the Druids, I leave to conjecture from the stories of our ancient Britons, who, had they left records of their prodigies in this kind, would doubtless have furnished us with examples as remarkable for the growth and stature of trees as any which we have deduced from the writers of foreign countries; since the remains of what are yet in being (notwithstanding the havoc which has universally been made, and the little care to improve our woods), may stand in fair competition with any thing that antiquity can produce."

Two species of Oak are indigenous to Britain, and they have been named by Botanists *Quercus robur*, and *Quercus sessiliflora*. The name *Quercus* is derived from the Celtic "*quer*," beautiful, and *cuez*, a tree. *Robur*, according to some, is derived from the Latin *robur*, strength; but we may, with greater probability, trace it to the Celtic *rove*, another name for the Oak, whence the Latins obtained their name for the tree, and subsequently adopted the same word to express the abstract idea of strength.* The name *Quercus robur*, therefore, rendered into English, means, "the tree of beauty and strength." *Quercus sessiliflora* is distinguished from the first species by having its fruit almost *sessile*, or *sitting* in groups on the leafy twig, without the intervention of any proper stalk; whilst *Quercus robur*, or *Quercus pedunculata*, as it is sometimes called, bears its fruit two or three together on a long peduncle, or fruit-stalk. But as this distinction is a modern one, and belongs rather to the naturalist than to

* In the north of Italy, the Oak is still called *Rovore*.

the poet or the historian, the names *Quercus robur* and *Oak*, when met with in English books not of a scientific character, must be understood to include both species. The word Oak is identical with the Saxon *aack* or *ak;* from which, also, *acorn* is derived. Hence Turner, the earliest English author on this subject, says; " Oke, whose fruite we call an acorn or an eykorn, (that is, y[e] corn or fruite of an Eike,) are harde of digestion and norishe very much, but they make raw humores. Wherefore we forbid the use of them for meates."

But finally, not to weary you with etymologies, when you expected to read about trees and woods;—from the Celtic *derw*, an Oak, the Druids took their name; and hence, also, the Greeks called the tree *drys*, and gave the appellation of *Dryads* to the imaginary beings who peopled their woods.

Like most long-lived trees, the Oak is of slow growth, averaging about a foot and a half in circumference in twenty years, and increasing about one inch in a year for the next century of its existence; after which its rate of growth diminishes. The extreme slowness of this increase may be better estimated by contrasting it with that of the Larch, which is very rapid in its formation of timber. An Oak at Wimbush, in Essex, in thirteen years had increased four inches and a half in circumference; and in the same time a Larch had increased thirty-three inches, or nearly eight times as much. The Oak does not usually attain any great height, being more remarkable for the thickness of its bole, and its widely-spreading head. Exceptions, however, are not wanting. In the Duke of Portland's park, at Welbeck, there stood, in 1790, an Oak, called "the Duke's walking-stick," which was an hundred and eleven feet high, the trunk rising to the height of seventy feet before it formed a head. Others nearly equalling this have been noticed.

A remarkable characteristic of the Oak is the stoutness of its limbs. "We know no tree, except, perhaps, the Cedar of Lebanon, so remarkable in this respect. The limbs of most trees spring from the trunk: in the Oak they may be rather said to divide from it; for they generally carry with them a great share of the substance of the stem: you often scarcely know which is stem and which is branch; and, towards the top, the stem is entirely lost in the branches. This gives peculiar propriety to the epithet 'fortes,' in characterising the branches of the Oak; and hence its sinewy elbows are of such peculiar use in ship-building. Whoever, therefore, does not mark the *fortes ramos* of the Oak, might as well, in painting a Hercules, omit his muscles. But I speak only of the hardy veterans of the forest. In the effeminate nurslings of the grove we have not this appearance. There the tree is all stem drawn up into height. When we characterise a tree, we consider it, in its natural state, insulated, and without any lateral pressure. In a forest, trees naturally grow in that manner. The seniors depress all the juniors that attempt to rise near them; but in a planted grove all grow up together, and none can exert any power over another.

"The next characteristic of the Oak is the twisting of its branches. Examine the Ash, the Elm, the Beech, or almost any other tree, and you may observe in what direct and straight lines the branches in each shoot from the stem; whereas the limbs of an Oak are continually twisting here and there in various contortions, and, like the course of a river, sport and play in every possible direction, sometimes in long reaches, and sometimes in shorter elbows."

"Another peculiarity of the Oak is its expansive spread. This, indeed, is a just characteristic of the Oak; for its boughs, however twisted, continually take a horizontal direction, and overshadow a large space of ground. Indeed, where it is fond of its situation, and has room to spread, it extends itself beyond any other tree, and, like a monarch, takes possession of the soil. The last characteristic of the Oak is its longevity, which extends beyond that of any other tree; perhaps the Yew may be an exception. I mention the circumstance of its longevity, as it is that which renders it so singularly picturesque. It is through age that the Oak acquires its greatest beauty, which often continues increasing even into decay, if any proportion exist between the stem and the branches. When the branches rot away, and the forlorn trunk is left alone, the tree is in its decrepitude in the last stage of life, and all beauty is gone."

Gilpin concludes this characteristic description with the following words: "I have dwelt the longer on the Oak, as it is confessedly both the most picturesque tree in itself, and the most accommodating in composition. It refuses no subject either in natural or in artificial landscape. It is suited to the grandest, and may with propriety be introduced into the most pastoral. It adds new dignity to the ruined tower and Gothic arch; by stretching its wild, moss-grown branches athwart their ivied walls, it gives them a kind of majesty coeval with itself; at the same time its propriety is still preserved, if it throw its arms over the purling brook, or the mantling pool, where it beholds

'Its reverend image in th' expanse below.'"

The diameter of the trunk of the Oak where it first leaves the ground, is generally much greater than it is a few feet higher. To this circumstance, and to the fact that its roots are not nearly so liable to rot in the ground as those of other trees, it may be attributed that it is very rarely blown up by the roots. That ingenious mechanic, Mr. Smeaton, is stated to have taken his idea of the form of the Eddystone Lighthouse from observing the proportions of an Oak trunk. Britton, in his "Beauties of Devon," thus writes: "The object from which Mr. Smeaton conceived his idea of rebuilding the Eddystone Lighthouse was the waist or bole of a large spreading Oak, which, though subject to a very great impulse from the agitation of violent winds, resists them all, partly from its elasticity, and partly from its natural strength. Considering the particular figure of the tree, as connected with its roots, which lie hid below ground, Mr. Smeaton observed that it rose from the surface with a large swelling base, which at the height of its own diameter is generally reduced by an elegant curve, concave to the eye, to a diameter less by at least one-third, and sometimes to half its original base. From thence its sides, tapering more gradually, assume a perpendicular direction, and for some height form a cylinder. After that a greater circumference becomes necessary for the insertion and establishment of the principal boughs, which produce a swelling of its diameter. Hence may be deduced an idea of what the proper shape of a column of the greatest stability ought to be to resist the action of external violence, when the quantity of matter is given whereof it is to be composed. Upon this model, therefore, on the 25th of August, 1759, Mr. Smeaton completed his lighthouse, being the third structure of the kind which had been raised on the dangerous rock from which it derives its name." How wisely he acted in choosing Nature for his instructress, may be inferred from the fact that it has now stood eighty-six years without requiring any essential repairs.

The trunk of the Oak, thus perfectly adapted as it is by its form to resist the most violent action of the wind, derives additional strength from the slow rate of growth of its timber. A very small quantity of woody fibre is deposited every year, but it is proportionately dense and solid, and the concentric annual layers are very firmly united. Hence it is admirably prepared to withstand lateral violence, as well as to support its enormous super-incumbent weight of branches; while its tap-root, descending perpendicularly to a great depth, and its tortuous underground arms proceeding horizontally at a greater depth beneath the surface than those of most other trees, are equally efficacious in resisting any upheaving force to which its spreading and abundant foliage might otherwise render it peculiarly liable.

Were it not for this wonderfully massive structure of the main trunk, the Oak would be unable to bear up the ponderous weight of its enormous limbs, which, each a mighty tree in itself, would rend in pieces any less substantial support. For it must have been remarked by every one who has looked thoughtfully on a full-grown Oak, that the trunk does not divide into several smaller ones, all approaching to a perpendicular direction; but that its unwieldy arms quit the bole almost horizontally, so that the centre of gravity of each lies a long way without the base of the tree, and is therefore constantly exerting its utmost power to tear itself away from the central column. This tendency to preserve a horizontal direction is most conspicuous in a full-grown tree, owing to the greater size of the object. But this peculiarity has not escaped the curious eye of the artist even in the smallest twigs. "In the spray of trees," Gilpin remarks, "Nature seems to observe one simple principle; which is, that the mode of growth in the spray corresponds exactly with that of the larger branches, of which indeed the spray is the origin. Thus the Oak divides his boughs from the stem more horizontally than most other deciduous trees: the spray makes exactly, in miniature, the same appearance; it breaks out in right-angles, or in angles that are nearly so, forming its shoots commonly in short lines, the second year's shoot usually taking some direction contrary to that of the first. Thus the rudiments are laid of that abrupt mode of ramification for which the Oak is so remarkable. When two shoots spring from the same knot, they are commonly of unequal length; and one with large strides generally takes the lead. Very often, also, three shoots, and sometimes four, spring from the same knot. Hence the spray of this tree becomes thick, close, and interwoven; so that at a little distance it has a full, rich appearance, and more of the picturesque roughness than we observe in the spray of any other tree. The spray of the Oak generally springs from the upper, or the lateral parts of the bough; and it is this which gives its branches that horizontal appearance which they generally assume."

This characteristic, which renders the Oak so great a favourite with the painter, makes it no less serviceable to the ship-builder, who selects the crooked limbs, and applies them, under the designation of *knee-timber*, to the purpose of forming the ribs of ships. Trees which grow at a considerable distance from each other are the most productive of this kind of timber; for, thus situated, the branches have ample room to follow the direction of the straggling roots, to which they naturally incline. In some parts of France, it is said, young trees are forced to assume this curved mode of growth by the suspension of weights to their heads; and in this country also experiments have been tried in order to produce similar results; but in most cases with very doubtful success. This custom was known to Virgil.

> "Continuo in sylvis magnâ vi flexa domatur
> In burim, et curvi formam accipit ulmus aratri."
>
> *Georg.* ii.

Evelyn says, "I conclude with recommending the bowing and bending of young timber-trees, especially Oak and Ash, into various flexures, curbs, and postures, which may be done by humbling and binding them down with tough bands and withs, or hooks rather, cut skrew-wise, or slightly haggled and indented with a knife, and so skrewed into the ground, or by hanging of weighty stones to the tops or branches, till the tenor of the sap, and custom of being so constrained, do render them apt to grow so of themselves, without power of redressing. This course would wonderfully accommodate the ship-builder with materials for knee-timber, and prove useful to the wheel-wright, as it would conform the wood to their moulds, save infinite labour, and abbreviate the work of hewing and waste."

According to the same author, (who is evidently a truthful writer, although many of his opinions appear to us very absurd, owing to the imperfect state of science in his time,) the Oak owed much of its popularity to the belief that its shade was remarkably salubrious. He says, "It is reported that the very shade of this tree is so wholesome, that the sleeping or lying under it becomes a present remedy to paralytics, and recovers those whom the mistaken malign influence of the Walnut-tree has smitten. The antients, who were fond of refreshing themselves under the shade of trees, caution us against the influence of the Walnut."

The foliage of the Oak is as characteristic as any other feature of the tree, whether we regard the sinuated form of each individual leaf, or the aggregate tufts. The principal difference between the leaves of *Quercus pedunculata* (or *Quercus robur*,) and *Quercus sessiliflora* is, that in the former they have scarcely any stems, whereas the leaves of the latter are decidedly stalked, and the lobes on each side are more nearly opposite. Both species burst their leaf and flower-buds about the same time, in April or May; *Quercus sessiliflora* being, however, generally somewhat later. At this time their pale green tint, delicately shaded with crimson, seems scarcely to accord with the bulky and more robust character of the rest of the tree; but, as the season advances, they assume a full, florid green, which they retain till very late in the year. At the approach of winter they put on a rich russet brown or red hue, and light up many a landscape, which without them would be cold and cheerless. Young trees do not cast their leaves even when every semblance of life has departed from them, but retain them, probably, as a protection for the embryo buds of the succeeding year, which are formed many months before they begin to expand. Once, on a frosty morning in January, I happened to be starting on a botanical ramble, and, just as the sun rose, entered a Devonshire lane, the hedges of which were topped with young Oaks laden with the last year's foliage. Suddenly it seemed as if I had been set down in Aladdin's wonderful garden. The trees, as they caught the first beams of the sun, appeared no longer to bear leaves, but plates of crimson transparent metal, or flakes of fire. The illusion lasted only a few minutes; for as soon as the sun was high enough to shine down *upon* the leaves, and not *under* them, they became withered oak-leaves again, bringing back to the mind the year that was past, with its cares and its blessings. This was one of those trifling incidents in a man's life, which, impressive in spite of its unimportance, *will* be remembered. *Being* remembered, it serves to illustrate my subject, and gives me the opportunity of saying that the field Naturalist has many such fertile reminiscences to fall back upon in his moments of reflection; more, perhaps, than any other man. This incident in particular reminds me not only of a happy day spent in the woods, but, besides this, it traces much more satisfactorily than any historian ever did, the worship of my Pagan forefathers to its source. Might it not have been under the influence of feelings such as were then excited in me, that the first Celt who worshipped Tornawa, the god of thunder, under the form of an Oak, received his primary impressions of the Divine origin of Nature? It were well if we, whenever we saw the symbol of God in His works, were to imitate the devotion of such an one, instead of resting merely on "poetic beauty." Such light as he had, he followed. To what might not *we* ascend if we as faithfully suffered ourselves to be guided by the Sun of Revelation?

The Oak is remarkable for sending out young shoots of spring foliage late in the season, when its proper leaves are fully matured; and this is more particularly the case when the latter have been injured. On the 2nd of August, 1844, the exposed Oaks at Penrose in Cornwall suffered severely from a violent storm from the west. In the course of a few hours all the leaves which had been unprotected from its influence, shrivelled up, (without, however, acquiring the true autumnal tint,) and died. But not long after, a second spring, as it were, set in, and the trees were partially restored to their former flourishing condition. White, noticing a similar occurrence, says: "When Oaks are quite stripped of their leaves by chaffers, they are clothed again, soon after Midsummer, with a beautiful foliage; but Beeches, Horse-chestnuts, and Maples, once defaced by those insects, never recover their beauty again for the whole season."

Amongst the many remarkable trees in the New Forest in Hampshire, is one called the Cadenham Oak, which buds every year in the depth of winter. Gilpin says, "Having often heard of this Oak, I took a ride to see it on the 29th of December, 1781. It was pointed out to me among several other Oaks, surrounded by a little forest stream, winding round a knoll on which they stood. It is a tall straight plant, of no great age, and apparently vigorous, except that its top has been injured, from which several branches issue in the form of pollard-shoots. It was entirely bare of leaves, as far as I could discern, when I saw it, and undistinguishable from the other Oaks in its neighbourhood; except that its bark seemed rather smoother, occasioned, I apprehended, only by frequent climbing. Having had the account of its early budding confirmed on the spot, I engaged one Michael Lawrence, who kept the White Hart, a small alehouse in the neighbourhood, to send me some of the leaves to Vicar's Hill, as soon as they should appear. The man, who had not the least doubt about the matter, kept his word, and sent me several twigs, on the morning of the 5th of January, 1782, a few hours after they had been gathered. The leaves were fairly expanded, and about an inch in length. From some of the buds two leaves had unsheathed themselves, but in general only one. One of its progeny, which grew in the gardens at Bulstrode, had its flower-buds perfectly formed so early as the 21st of December, 1781.

"This early spring, however, of the Cadenham Oak, is of very short duration. The buds, after unfolding themselves, make no further progress, but immediately shrink from the season, and die. The tree continues torpid, like other deciduous trees, during the remainder of the winter, and vegetates again in the spring, at the usual season. I have seen it in full leaf in the middle of summer, when it appeared, both in its form and foliage, exactly like other Oaks." *

Dean Wren, speaking of this tree, says, "King James could not be induced to believe the τὸ ὅτι (*reason*) of this, till Bishop Andrewes, in whose diocese the tree grew, caused one of his own chaplaines, a man of known integritye, to give a true information of itt, which he did: for upon the eve of the Nativitye he gathered about a hundred slips, with the leaves newly opened, which he stuck in claye in the bottom of long white boxes, and soe sent them post to the courte, where they deservedly raised not only admiration, but stopt the mouth of infidelitye and contradiction for ever. Of this I was both an eye-witness, and did distribute many of them to the great persons of both sexes

* A writer in the Saturday Magazine explains this phenomenon, on the supposition that the tree was originally brought by some enthusiastic pilgrim from the Holy Land, and continued to put forth its leaves at the same season that it had budded in Palestine. This supposition is undoubtedly very ingenious; but, unfortunately, the British Oak does not grow in Palestine, nor any other species so closely resembling it as to be easily confounded with it.

in court and others, ecclesiastical persons. But in these last troublesome times, a divelish fellow (of Herostratus humour) having hewen itt round at the roote, made his last stroke on his own legg, whereof he died, together with the old wondrous tree: which now sproutes up againe, and may renew his oakye age againe, iff some such envious chance doe not hinder or prevent itt; from which the example of the former villane may perchance deterr the attempte. This I thought to testifie to all future times, and therefore subscribe with the same hand through which those little oakye slips past."

In many of the rural districts oak-leaves and oak-apples (to be mentioned hereafter) are worn by boys on the 29th of May, the anniversary of the Restoration of Charles II., who is said to have concealed himself in an Oak tree from the Parliamentary soldiers.*

I must not omit to mention here that the Romans were accustomed to bestow a wreath composed of Oak leaves, called a civic crown, on any one who saved the life of a citizen; which was considered the highest service that could be rendered to the state.

"An oaken wreath his hardy temples bore,
Mark of a citizen preserved he wore."

ROWE'S *Lucan.*

Here, too, I may mention the absurd belief, once popularly prevalent, that the Barnacle-goose owed its origin to this tree. The word *barnacle* is said to be derived from *bairn*, a child, and *acle*, the *aac*, or oak. The quaint old botanist, Gerard,

* For a full account of King Charles's Oak see page 83.

tells the story so faithfully, that I cannot do better than transcribe his own words. "There are found in the North of Scotland, and islands adjacent, called Orchades, certain trees whereon do grow certain shells tending to russet, wherein are contained little living creatures; which shells, in time of maturitie, do open, and out of them do grow those little living things, which, falling into the water, do become fowles, which we call *barnakles;* in the North of England, *brent-geese;* and in Lancashire, *tree-geese;* but the other that do fall upon the land perish, and come to nothing. Thus much from the writings of others, and also from the mouths of people of those parts, which may very well accord with truth." This he gives from the report of others; now for what is proved by the evidence of his own senses. "There is a small island in Lancashire, called the Pile of Toulders, wherein are found the broken pieces of old and bruised ships, some whereof have been cast there by shipwracke; and also the trunks and bodies, with the branches, of old and rotten trees, cast up there likewise, whereon is found a certain spawn, or froth, that in time breaketh into certain shells, in shape like those of the muskle, but sharper pointed, and of a whitish colour, wherein is contained a thing in form like a lace of silke, finely woven as it were together, of a whitish colour, one end whereof is fastened unto the inside of the shell, even as the fish of oisters and muskles; the other end is made fast unto the belly of a rude mass, or lumpe, which in time cometh to the shape and form of a bird. When it is perfectly formed the shell gapeth open, and the first thing that appeareth is the foresaid lace, or string;

next come the legs of the bird hanging out; and as it groweth greater, it openeth the shell by degrees, till at length it is all come forth, and hangeth only by the bill; in short space after it cometh to full maturitie, and falleth into the sea, where it gathereth feathers, and groweth to a fowl bigger than a mallard, and lesser than a goose, having black legs, bill, or beake, and feathers black and white, spotted in such a manner as our magpie; called in some places a *pie-annet;* which the people of Lancashire call by no other name than a *tree-goose;* which place aforesaid, and the parts adjoining, do much abound therewith that one of the best is bought for three-halfpence. For the truth hereof, if any doubt, let them repaire to me, and I shall satisfie them by the testimonie of good witnesses."*

This strange fable took its rise from a certain shell-fish being frequently found attached to pieces of wood which had long lain in salt-water. This shell-fish, now called *Lepas anatifera,* is provided with a long leathery tube, by which it attaches itself to the bottom of vessels, and to other timber; it is also furnished near the other extremity with a number of curved, feathery fibres, which, when expanded, bear some resemblance to the tail of a bird.† From this fancied similarity, and the

* Herball, p. 1588.

† "It is hardly worth while to mention the *clayks*, a sort of geese, which are believed by some, with great admiration, to grow upon trees on this coast, and in other places; and, when they are ripe, to fall down into the sea, because neither their nests nor eggs can anywhere be found. But they who saw the ship in which Sir Francis Drake sailed round the world, when it was laid up in the river Thames, could testify that little birds bred in the old rotten keels of ships, since a great number of such, without life and feathers, stuck close to the outside of the keel of that ship. Yet I should think that the ge-

coincidence that the shell-fish was found in abundance in places which the Barnacle-goose frequented, probably to make them its food, the fable originated; a fertile imagination making up for the barrenness of the facts. Before the Reformation, Dr. Walsh tells us, the fishy origin of the bird was so firmly believed that the question was warmly and *learnedly* disputed, whether it might not be eaten in Lent.

The story may have gained a more ready credence from the fact that the Oak is more prolific in animal life, supplying more insects with food, than any other tree. According to Mr. Stephens, an excellent authority, nearly a half of the British insects which feed on vegetables, either exclusively or partially inhabit the Oak. If to this number we add the insects which live on the above, it will be found that the total of insects which, during some period of their existence, derive their support either from the tree itself, or from their fellow-colonists in it, will amount to scarcely less than two thousand. Of these I shall mention a few of the most remarkable, referring the reader who is anxious to learn further particulars to an able article by Westwood in Loudon's "Arboretum Britannicum," a work of great research, abounding in valuable information on all subjects connected with the history, propagation, and treatment of all the trees and shrubs, both indigenous and exotic, growing in Great Britain.

Among those insects which feed on the substance of the wood of the Oak, is the larva or grub

neration of these birds was not from the logs of wood, but from the sea, termed by the poets "the parent of all things."—*Camden's Britannia.*

of the great Stag-beetle.* Notwithstanding, however, its abundance in some localities, and its great size, it does no injury, never attacking any but decayed wood. When it has attained its full

size, it constructs a cocoon of chips of wood, which it glues together by a self-derived cement, and assumes the pupa stage of its existence, when it ceases to eat. The perfect insect, well known as the Stag-beetle, seems to subsist entirely upon fluids, which it laps up by means of its long, pencil-like lower jaws and lip. The number of the insects which feed upon the living wood appears to be limited; but those which reside beneath the bark, without boring into the wood, are much more numerous. So great are the ravages sometimes committed by one minute beetle (*Scolytus pygmæus*), that it was, not long since, found necessary to cut down, in the Bois de Vincennes near Paris, 50,000 young Oaks in which they had

* Lucanus cervus.

taken up their abode. It is, however, from the leaves of the Oak that the chief portion of its insect population derive their support; and it is principally amongst the caterpillars of the moths and butterflies that the greatest number of the leaf-feeders are found. "Of these, the Tortrix viridana, a very small, pretty, green species, is by far the most obnoxious; entirely stripping the Oaks of their foliage, as we have more than once observed at Coombe Wood in Surrey. Even the smaller sorts of caterpillars become, from their multiplicity, sometimes as destructive as those which are of considerable magnitude. During the summer of 1827, we were told that an extraordinary blight had suddenly destroyed the leaves of all the trees in the Oak of Honour Wood, Kent. On going thither, we found the report but little exaggerated; for, though it was in the leafy month of June, there was scarcely a leaf to be seen on the Oak trees, which constitute the greater portion of the wood. But we were rather surprised when we discovered, on examination, that this extensive destruction had been effected by one of the small solitary leaf-rollers (*Tortrix viridana*): for one of this sort seldom consumes more than four or five leaves, if so much, during its existence. The number, therefore, of these caterpillars must have been almost beyond conception; and that of the moths, the previous year, must also have been very great; for the mother moth only lays from fifty to a hundred eggs, which are glued to an oak branch, and remain during the winter. It is remarkable that in this wood, during the two following summers, these caterpillars did not abound. The moth varies in the expansion of its wings from seven to thirteen lines: the anterior wings are pale green, with a whitish margin in front, and the posterior wings brownish. It is so extremely abundant, that towards the end of

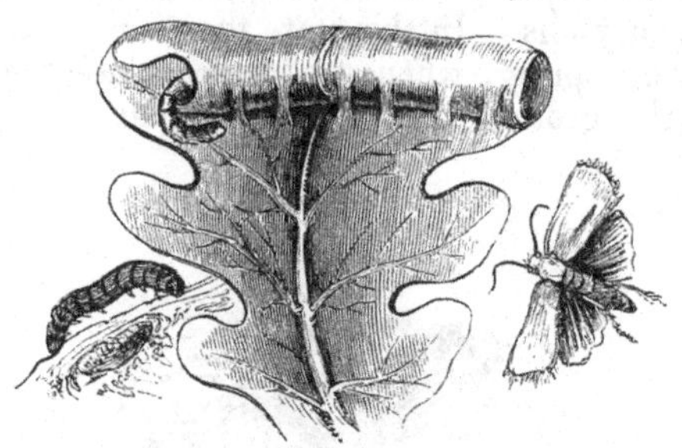

the month of June, when it first appears, it may be shaken from the trees in perfect showers. The caterpillar of this moth rolls up the oak-leaves in a very ingenious manner, so as to form a very commodious retreat, in which, indeed, it ordinarily resides, the centre of the roll being open: its diameter is proportionate to that of the body of the insect, and the roll is secured by various little packets of silk attached to the body of the leaf and to the adjoining part of the roll."*

Among the beetles, the common Cockchafer, or Oakweb,† "is the most obnoxious of the leaf-eating species. The egg of this terrible devastator is white, and is deposited in the ground, where it soon changes into a soft whitish grub, with a red head, and about an inch and a half long. In this state it continues four years, during which time it

* Loudon, Arbor. Brit., cap. cv. † Melolontha vulgaris.

commits the most destructive ravages on the roots not only of grass, but of all other plants and young trees. When full grown, the larvæ dig in the earth to the almost incredible depth of five or six feet, spin a smooth case, and then change into a chrysalis. In this state they remain till the following spring, when the perfect insect comes from the ground, and commences an immediate

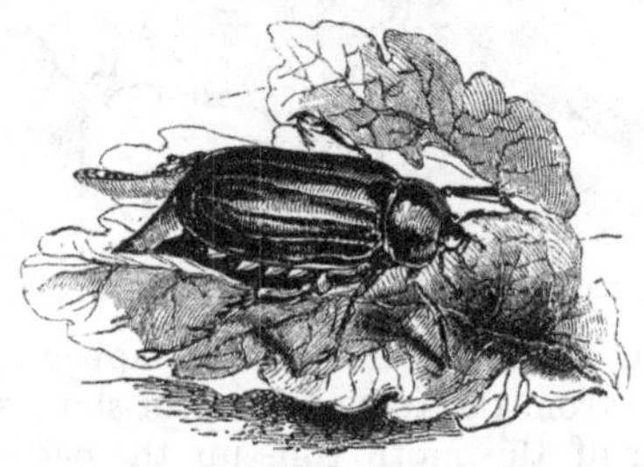

attack on the leaves of trees. A remarkable account of the ravages of these insects is given by Molyneux, in one of the early volumes of the Philosophical Transactions, in which their appearance in the county of Galway in Ireland, in 1688, is narrated. They were seen in the daytime perfectly quiet, and hanging from the boughs in clusters of thousands, clinging to each other like bees when they swarm; but dispersing towards sunset, with a strange humming noise, like the beating of distant drums; and in such vast numbers, that they darkened the air for the space of two or three miles square; and the noise they made in devouring the leaves was so great, as to resemble the distant sawing of timber. In a very short time the leaves of all the forest-trees for some miles were destroyed, leaving the trees as bare and desolate in the middle of summer as they would have been in winter; they also entered the gardens, and attacked the fruit-trees in the same manner. Their multitudes spread so exceedingly, that they infested houses, and became exceedingly offensive and troublesome. They were greedily devoured by the swine and poultry, which watched under the trees for their falling, and became fat on this unusual food; even the people adopted a mode of dressing them, and used them as food. Towards the end of the summer they disappeared suddenly, and no traces were perceived of them the ensuing year. In the Magazine of Natural History, a story is told of a gentleman, who, finding his Oak trees stripped of their leaves in the middle of summer, suspected some rooks of having destroyed them. That the Oaks were nearly bare was beyond dispute; and he had himself seen the rooks settling on them, and pecking away right and left with their bills. War was, therefore, declared against the rooks; but, fortunately, before hostilities were commenced, the gentleman was convinced, by some one who knew more of natural history than himself, that the rooks were not in fault: on the contrary, they had only flocked to the trees for the sake of devouring the myriads of Cockchafers, and of the larvæ of Moths, which were the real depredators."*

Among the less injurious insects which frequent the Oak is the Purple Emperor, the most

* Loudon, Arbor. Brit., cap. cv. The service performed by the rook in destroying the grub of the Cockchafer is well known.

splendid of the British butterflies. The caterpillar of this insect feeds on the leaves of one of our Willows: but the perfect insect, according to Haworth, "invariably fixes his throne upon the summit of a Broad-leaved Oak, from the utmost sprigs of which, on sunny days, he performs his aerial excursions; and in these ascends to a much greater elevation than any other insect I have ever seen, sometimes mounting higher than the eye can follow; especially if he happens to quarrel with another Emperor, the monarch of some neighbouring Oak: they never meet without a battle, flying upwards all the while, and combating with each other as much as possible; after which they will frequently return again to the identical sprigs from which they ascended."

To the class of innocuous insects must be referred also the various species of gall-flies, whose instinct teaches them to originate a local disease in some part of the Oak, and thus to provide their offspring with food and a dwelling-house. A history of the Oak would be very imperfect without a full notice of the curious productions known by the name of galls; and, as the subject is a very interesting one, I do not scruple to dwell upon it, although, strictly speaking, it belongs as much to Entomology as to Botany.

A small fly alights on a twig, or leaf, or bud, of an Oak, and with an excessively acute instrument, with which it is provided by nature for this express purpose, punctures the vegetable fibre, and deposits an egg, or perhaps two or more eggs, so minute as to be almost invisible to the human eye. And is this all the provision that the fly is going to make for its progeny? It is; for though seemingly little, it is amply enough. The preservation of that egg is the care of the Almighty. You believe that your own life is under the protection of the same beneficent Being; and yet the

egg of a fly, which is so small that you might brush away a thousand of them with the palm of your hand, bears an infinitely greater proportion to the world which you think so large, than the whole term of your life does to eternity. I do not wish to place the value of an immortal soul in the scale against any material object; it would be wicked to do so, for it would be to set aside the beautiful lesson of our blessed Redeemer; but I would warn you against the habit of believing that the preservation of any created thing (however contemptible it may appear to you) is too trifling a matter for the care of the Almighty. And here, too, it is particularly necessary that I should advert to the all-fostering Providence of God, because the deepest and most learned speculations of human science are utterly at fault. Why from the puncture of one kind of fly a large irregular excrescence should be produced; why from that of another a smooth spherical gall, or a scaly bud, or a flat circular scale, is all a mystery,—a mystery so deep that no plausible explanation of it has ever been attempted. To say that an alteration takes place in the character of the juices; that a disease is produced which arrests them, and causes them to arrange themselves in a certain set form—this is not to account for the phenomenon; it is merely an unsatisfactory statement of the result, the real difficulty being left untouched. You must therefore be content to read the description of the different kinds of galls which have been observed, and test its accuracy, when you can, by comparing it with the natural objects themselves.

In the first place, it appears that the different kinds of insects select different parts of the tree in which to deposit their eggs, and that the character of the galls produced equally varies. The largest species is generally called the Oak-apple, and grows on the extremity of a twig. It is of a soft spongy substance, and an irregular shape, shaded with brown and pink on the outside; and it is divided on the inside into a number of cells, each of which contains either a small grub, a pupa, or a perfect fly, according to the season. Gerard, whose marvellous account of the barnacle-goose I have already quoted, tells us, that galls of this kind were in his day commonly consulted as auguries. "The oke-apples," he says, "being broken in sunder about the time of their withering, doe foreshew the sequell of the yeare, as the expert Kentish husbandmen have observed by the living things found in them; as, if they finde an ant, they foretell plenty of graine to ensue; if a white worm, like a gentill or magot, then they prognosticate murren of beasts and cattell; if a spider, then (say they) we shall have a pestilence, or some such like sicknesse to follow amongst men. These things the learned also have observed and noted; for Matthiolus, writing upon Dioscorides, saith, that before they have an hole through them they containe in them either a flie, a spider, or a worme; if a flie, then warre ensueth; if a creeping worme, then scarcity of victuals; if a running spider, then followeth great sickenesse or mortalitie." *

* "The presage of the year preceding, which is commonly made from insects or little animals in oak-apples, according to the kinds thereof, either maggot, fly, or spider: that is, of famine, war, or pestilence; whether we mean that woody excrescence, which shooteth from the branch about May, or that round and apple-like accretion which groweth under the leaf about the latter end of summer, is, I doubt, too distinct, nor verifiable from event. For flies and maggots are found every year, very seldom spiders: and Helmont affirmeth, he could never find the spider and the fly upon the same trees, that is the signs of war and pestilence, which often go together: besides, that the flies found were at first maggots, experience hath informed us; for keeping those excrescences, we have observed their conversions, beholding in magnifying glasses the daily progression thereof. As may be also observed in other vegetable excretions, whose maggots do terminate in flies of constant shapes; as in the nut-galls of the outlandish Oak, and the mossy tuft of the wild briar; which having gathered in November, we have found the little maggots, which lodged in wooden cells all winter, to turn into flies in June."—SIR T. BROWNE'S *Vulgar Errors*.

In this case, as in many others, truth is stranger than fiction. It not unfrequently happens that one of the ichneumon flies lays an egg in the body of the original inhabitant of one of these cells. From this egg proceeds a small worm, which lives on the substance of its predecessor, inhabits his house, and, when grown to a perfect insect, escapes and takes flight in search of a similar abode for its own progeny. What faculty, or sense, or instinct, can this little animal possess, which directs it to a solid vegetable substance, in the centre of which is stored up proper nourishment for its young? What geometrical skill enables it to discover in what part of the mass its prey lies buried? By the aid of what calculating power does it contrive to pierce the body of the included grub only so deep as to deposit its egg in a place of security, without wounding any vital part? If old Gerard had been acquainted with these facts, he would surely have thought them more wonderful than all his absurdly superstitious fancies, which he could scarcely have been so foolish as to believe when he wrote them down.

The most remarkable kind of oak-gall, next to that described, is produced by another insect of the same genus. (Cynips.) This fly deposits its eggs in the stalk of the stamen-bearing flowers, which is long and drooping. The excrescence which follows resembles a currant in size, shape, and even in mode of growth, it often happening that several are placed at short distances from each other on the same thread-like stem. There is a remarkable fact connected with this species of gall. Those flowers of the Oak which bear stamens only are destined to wither and fall off as soon as they have shed their pollen, being no longer of any use. Those stalks, however, to which galls are attached, remain firmly united with the tree, and grow vigorously as long as the

grubs contained in them continue to feed. The notice of this circumstance has helped to decide one of the controverted points in vegetable physiology,—namely, whether the sap is forced into motion by some power residing within the tree, or whether, being evaporated or otherwise consumed at the extremities, a fresh supply rises to fill its place. In favour of the latter supposition many arguments may be adduced; and, among them, that furnished by the flower-stalks of the Oak, which cease to receive sap, and wither away, as soon as their proper vital functions are at an end; but if they bear swelling galls on their extremities, contribute their share of influence towards the ascent of sap throughout the tree generally, and supply their own substance in particular with as much as is required.

Another gall, resembling the last in form, being spherical, is found attached to the *leaves* of the

Oak. These vary very much in size, some being as large as a marble; and each contains a single

insect, which, when it arrives at its perfect state, eats its way out through a great portion of the solid substance of the gall.

The habitation of all the parasitic insects hitherto mentioned is formed out of the pulpy substance of the tree: one, however, which is not uncommon, and is called the Artichoke gall, is an irregular development of the bud, and consists of a number of leafy scales overlapping each other. At first sight it might almost be taken for a young cone; but on dissection is found, like other galls, to contain insects in various stages of their growth according to the season.

Another singular appendage of the leaf is the Oak-spangle, a flat circular disc, attached by its central point to the under-surface of the leaf. The inner side is smooth; the outer red, hairy, and fringed. Each of these contains a single insect, which retains its habitation until March, long after the leaves have fallen to the ground.

Another insect of the same genus (*Cynips*) deposits its eggs at the base of the trunk, immediately above the root. In the early Spring of the last year, 1845, I detected two galls formed by this species in Merthen Wood, Cornwall. The larger was about as big as a walnut, and produced in April sixty small flies, much resembling winged ants. They were not very active during their early existence, and possessed the remarkable instinct, common to many other insects, of counterfeiting death when touched.

The galls of commerce, I may here remark, are similar in their nature to those already mentioned. They are produced by a dwarf species of Oak (*Quercus infectoria*), which rarely attains the height of six feet, growing in Asia Minor and Persia. The insect which occasions this gall is of a pale colour, and may be often found in the galls sold in the shops of druggists. The latter vary greatly in the qualities on account of which they are employed; those which still contain the insect, and are known by the name of black, blue, or green galls, being the best: while those from which the insect has escaped, which are called white galls, do not contain more than two-thirds of the astringent qualities of the former. They are used for making ink, for dyeing, and for medicinal purposes.

Evelyn, who wrote in 1664, gives the following account of them. "Pliny affirms that the galls break out all together in one night, about the beginning of June, and arrive to their full growth in one day; this I should recommend to the experience of some extraordinary vigilant woodman, had we any of our Oaks that produced them, Italy and Spain being the nearest that do. Galls are of several kinds, but grow upon a different species of Robur from any of ours, which are never known to bring these excrescences to maturity; the white and imperforated are the best; of all which, and their several species, see Casp. Bauhinus, and the excellent Malphighius, in his Discourse de Gallis, and other morbous tumours raised by, and producing insects, infecting the leaves, stalks, and branches of this tree with a venomous liquor or froth, wherein they lay and deposite their eggs, which bore and perforate these excrescences when the worms are hatched, so as we see them in galls."

The apples of Sodom, or Dead Sea apples, described by Josephus as being beautiful to the eye, but composed internally of dust and bitter ashes, are by some recent authors, with much semblance of truth, considered to be galls of some species of Oak, containing insects.

I now come to speak of the flower and fruit of the Oak. Of the former, every tree produces two kinds; the first containing stamens only, and therefore producing no fruit. These appear nearly as soon as the leaves, consisting of yellow tasselled threads, which wither and drop off as soon as they have shed the pollen or fructifying dust, which they contain; unless, as I have stated above, they happen to have been perforated by one of the gall insects. The other kind of flower appears soon after, and is even less conspicuous than the first; it is this which subsequently produces the acorn. Of the acorn itself no description need be given; no other natural production, perhaps, has served as a model for so many ornamental works of art; and this is to be attributed not so much to the popularity of the Oak, as to the finished elegance of form of the fruit itself. 'Acorn-shaped' would, I should think, be a word as readily understood as 'round' or 'square.' Acorns and roses are in modern architecture what pomegranates and lilies were in Jewish. Different in proportions though it is in the various species of Oak, there is yet always similarity enough to detect the genus of the tree which produced it. The ball may be almost buried in the cup, or may be disproportionately long; the latter may be almost smooth, or rugged, or even mossy; yet, were an acorn of any species to be placed before a person who had never seen any other than that of the British Oak, he would immediately pronounce the tree from which it was gathered an Oak.

As an article of food, the acorn has been, and in many places still is, highly prized. In the time of Strabo, Rome was principally supplied with hogs which had been fattened on mast in the woods of Gaul. This mast is supposed to have included the acorns of the common and Turkey Oaks, and of the Ilex; as well as the nuts of the Beech and Chestnut. So important were acorns formerly considered, that by the laws of the Twelve Tables the owner of a tree might gather up his acorns though they should have fallen on another man's ground.

It appears from Domesday Book, that in England in the time of William the Conqueror, "Oaks were still esteemed, principally for the food they afforded to swine; for the value of the woods in several counties is estimated by the number of hogs they would fatten. The survey is taken so accurately that in some places woods are mentioned of a single hog. The numerous herds of swine which still continue one of the chief sources of wealth to the rural population of Spain, are fed on the acorns of the evergreen Oak, which abound in almost every part of the country. They are also a grateful food to deer, both when wandering at large in the forests and when confined in parks; and are greedily eaten by pheasants and partridges. Evelyn, recommending the extensive planting of Oaks, says, "In this poor territory (Westphalia) every farmer does by antient custom plant so many Oaks about his farm as may suffice to feed his swine. To effect this, they have been so careful, that when of late years the armies infested the poor country, (both Imperialists and Protestants,) the single bishoprick of Munster was able to pay one hundred thousand crowns *per mensem*, (which amounts to about twenty-five thousand pounds sterling, of our money,) besides the ordinary entertainment of their own princes and private families. This being incredible to be practised in a country so extremely barren, I thought fit to mention, either to encourage or reproach us." The same author says, that "a peck of acorns a-day, with a little bran, will make a hog, 'tis said, increase a pound weight *per diem* for two months together."

"The Rev. Mr. Robinson, in his Natural History of Westmoreland and Cumberland, says, that 'birds are natural planters of all sorts of trees, disseminating the kernels upon the earth till they grow up to their natural strength and perfection.' He tells us, that early one morning he observed 'a great number of rooks very busy at their work, upon a declining ground of a mossy surface, and that he went out of his way on purpose to view their labour. He then found that they were planting a grove of Oaks. The manner of their planting was thus: — they first made little holes in the earth with their bills, going about and about till the hole was deep enough, and then they dropped in the acorn, and covered it with earth and moss. The young plantation,' Mr. Robinson adds, 'is now growing up to a thick grove of Oaks, fit for use, and of height for the rooks to build their nests in. The season was the latter end of autumn, when all seeds are fully ripe.'" *

But the use of this fruit as an article of food is not confined to the inferior animals: even man has condescended to submit to the same humble fare, and among the rest our own progenitors. "The earliest notices which we have of the Oak in Britain are in the Saxon Chronicles, from which it appears that Oak forests were chiefly valued for the acorns which they produced, which were generally consumed by swine and other domestic animals, but, in years of great scarcity, were eaten by man. 'Famines,' Burnet observes, 'which of old so continually occurred, history in part attributes to the failure of these crops. Long after the introduction of Wheat and Oats, and Rye—nay, little more than seven hundred years since, when other food had in a great measure superseded the use of Mast, considerable reliance was still placed thereon, and Oaks were chiefly valued for the acorns they produced. In the Saxon Chronicles, that year of terrible dearth and mortality, 1116, is described as 'a very heavy-timed, vexatious, and destructive year,' and the

* Jesse's Gleanings in Natural History.

failure of the mast in that season is particularly recorded: 'This year, also, was so deficient in mast, that there never was heard such in all this land, or in Wales.'"*

During the Peninsular war, both the natives and the French frequently fed on the acorns they met with in the woods of Portugal and Spain. In Morocco and Algiers, the acorns of *Quercus ballota* are sold in the public markets, and eaten by the Moors both raw and roasted. Those of our own Oaks, when roasted, and treated like coffee, are said to afford a liquor closely resembling that beverage; and when sprouted acorns are treated like malt, they afford a liquor from which a strong spirit may be distilled. "Acorns," says Evelyn, "before the use of Wheat-corn was found out, were heretofore the food of men,—nay, of Jupiter himself, as well as other productions of the earth, till their luxurious palates were debauched. And even in the time of the Romans the custom was, in Spain, to make a second service of acorns and mast, as the French do now of marrons and chestnuts, which they likewise used to roast under the embers. And men had indeed hearts of oak; I mean not so hard, but health and strength, and lived naturally, and with things easily parable and plain. And even now I am told that those small young acorns which we find in the Stock-doves' craws are a delicious fare, as well as those incomparable sallads, young herbs taken out of the maws

* The Greeks, in allusion to the use of acorns as food, called one species of oak *phagos*, or *phegos*, and the Latins *esculus*, as much as to say, *the tree of eating*; like our word *mast* for acorns; whence *masten*, to feed or fatten, and *masticate*, to chew. From *glans*, the French derive their *glaner*, and we our *glean*, *gleaner*, for the collecting of scattered corn.—*Saturday Magazine.*

of Partridges at a certain season of the year, which gives them a preparation far exceeding all the art of cookery."

The acorns of the Balonia Oak (*Quercus ægilops*) are annually brought to England from the Levant and the Morea, and are in great demand for tanning; being said to contain more tannin in a given bulk of vegetable than any other substance.* The cups of this acorn are much larger than those of our British species, and are covered externally with long reflexed scales.

I have not yet spoken of the application of the various parts of the Oak to the arts of civilized life, it having been my object to devote as much of my space as possible to the tree in its natural state. But inasmuch as a notice of any tree, and especially this king of trees, would be of necessity considered imperfect without at least some few remarks on this head, I will proceed to give a brief history of the general uses to which the wood and other parts of the Oak may be applied.

The particular and most valued qualities of the Oak are hardness and toughness. Shakspeare used two epithets to express these qualities, which are perhaps stronger than any we can find.

> "Thou rather with thy sharp and sulph'rous bolt
> Split'st the *unwedgable* and *gnarled* Oak
> Than the soft myrtle."

"Many kinds of wood are *harder*, as box and ebony; many kinds are *tougher*, as yew and ash; but it is supposed that no species of wood, at least no species of timber, is possessed of both these qualities together in so great a degree as British Oak.

* The cups of this Oak, called "valonia," are now so extensively used, that Oak-bark has materially deteriorated in value.

Almost all arts and manufactures are indebted to it; but in ship-building, and bearing burdens, its elasticity and strength are applied to most advantage. I mention these mechanical uses only because some of its chief beauties are connected with them. Thus, it is not the erect, stately tree, that is always the most useful in ship-building; but more often the crooked one, forming short turns and elbows, which the shipwrights and carpenters commonly call knee-timber. This, too, is generally the most picturesque. Nor is it the straight, tall stem, whose fibres run in parallel lines, that is the most useful in bearing burdens; but that whose sinews are twisted, and spirally combined. This, too, is the most picturesque. Trees under these circumstances generally take the most pleasing forms." *

The admirable qualities of Oak as a material for building, and other purposes, were known to our ancestors in ages long past, scarcely any other timber being found in any buildings of very high antiquity. "The doors of the inner chapels of Westminster Abbey are said to be coeval with the original building: if this be true, they must be more than twelve hundred years old. The shrine of Edward the Confessor, in the same abbey, is also of Oak, and must be nearly eight hundred years old. In the county-hall at Winchester is preserved Arthur's round table, so well known in stories of English chivalry. It bears the figure of that Prince, and the names of several of his knights. Henry the Eighth is said to have taken great pleasure in shewing this table to his illustrious visitor, Charles the Fifth, as the actual oaken

* Gilpin's Forest Scenery.

table made and placed there by the renowned British Prince, Arthur, who lived in the early part of the sixth century; that is, about 1300 years ago. Hence the poet Drayton sings:—

> "And so great Arthur's seat ould Winchester prefers,
> Whose ould round table yet she vaunteth to be hers."

It must have been cut from a tree of immense girth, as it measures eighteen feet in diameter. It has been perforated in many places by bullets, supposed to have been shot by Cromwell's soldiers.

In digging away the foundation of the old Savoy Palace in London, which was built six hundred and fifty years since, the whole of the piles, many of which were of Oak, were found in a state of perfect soundness; as also was the planking which covered the pile-heads.* In clearing the channel at Brundisium, in Italy, "the workmen have drawn up many of the oak-piles that were driven in by Cæsar. They are small Oaks stripped of their bark, and still as fresh as if they had been cut only a month, though buried above eighteen centuries seven feet under the sand. These piles were driven in by Julius Cæsar to block up Pompey's fleet."†

Our forefathers appear to have discovered the eligibility of Oak as a material for ship-building at a very early period; the Alder, Cypress, and Pine, having been principally used by the Romans. Professor Burnet, writing on this subject, says, "An ancient vessel was discovered some years ago in a branch of the river Rother, near the west end of the Isle of Oxney, in Kent, and about two miles from the spot where formerly stood the Ro-

* Burnet. † Phillips.

man city of Anderida. The timber of which this vessel is constructed is Oak, perfectly sound, and nearly as hard as iron; and some persons believe it to be one of the fleet abandoned by the Danes after their defeat in the reign of Alfred. This, however, is but conjecture; still, whether it be so, or whether it be a wreck of some Danish pirates, it must have lain there many centuries. Sir Joseph Banks records, in the Journal of Science, the following account of an ancient canoe found in Lincolnshire in April, 1816, at a depth of eight feet under the surface, in cutting a drain parallel with the river Witham, about two miles east of Lincoln, between that city and Horsley Deep. It seems hollowed out of an Oak tree; it is thirty feet eight inches long, and measures three feet broad in the widest part. The thickness of the bottom is between seven and eight inches. Another similar canoe was discovered in cutting a drain near Horsley Deep; but it was unfortunately destroyed by the workmen before it was ascertained what it was. Its length was nearly the same as the other, but it was four and a half feet wide. Besides these, three other canoes, resembling the above in construction, have been found in the same county: one in a pasture near the river Trent, not far from Gainsborough; and two in cutting a drain through the fens below Lincoln. One of these is deposited in the British Museum. Conjecture alone can be indulged as to the probable age of these three canoes; but the fact of their being hollowed out of the trunks of old trees must carry them back to a very early date, and establish their extreme antiquity. Long before the time of Alfred the Britons were familiar with ships regularly built; vessels such as these are found only among the rudest people, and in the earliest stages of society; and the epoch when any of the European nations used such canoes must be remote indeed."

Speaking of the uses of the Oak generally, Loudon says, "The wood of the Oak is more durable, in every state in which it can be placed, than that of any other tree which abounds in large quantities in Europe. It is hard, tough, tolerably flexible, strong, without being too heavy, not easy to splinter, and not readily penetrated by water; and hence its value in ship-building. Some woods are harder, but they are more fragile; and others are more flexible, but do not possess so much hardness, toughness, and durability. Where the grain is twisted, no timber is so well adapted for posts, either in house-building or in setting up mills, engines, or large machines. No wood lasts longer where it is subject to be alternately wet and dry; and Oak piles have been known to endure many centuries. Shingles, poles and laths, last longer of this wood than of any other; and casks, and every other description of cooper's work, are most durable, and best adapted for containing wines, ales, and other liquors, when they are made of Oak. Oak timber is particularly esteemed for the spokes of wheels, for which the small and slow-growing Oak of mountainous districts is greatly preferred to the more rapid-growing and larger Oak of the valleys. Oaks of from fifteen to thirty years' growth make the most durable poles. The young tree, when from five to ten feet high, makes excellent hoops, which Evelyn says we ought to substitute for those of Hazel and Ash, as they are six times more durable: it also makes the very best walking-sticks, and very good handles to carters' whips. Of the roots, Evelyn says, were formerly made hafts to daggers, handles to knives, tobacco-boxes, mathematical instruments, tablets for artists to paint on instead of canvass, and elegant camleted joiner's work. Oak wood, every one knows, is preferred before all others for ship-building, in the temperate regions of both hemispheres. From its toughness, it does not splinter when it is struck by a cannon-ball, and the hole made by a ball is consequently easy to plug up. Throughout Europe, and more especially in Britain, Oak timber was used for every purpose, both of naval and civil architecture, till the wood of the pine and fir tribe came to be generally imported from the Baltic and North America, about the beginning of the last century. Since that period, the use of Oak timber has given way to that of pine and fir in house-building; but still it maintains its superiority in the construction of ships, and various kinds of machines; and even in house-building, where great durability is required. Oak wood is also still employed in joinery and cabinet-making."

A writer in the *Quarterly Review, Oct.* 1814, comparing the qualities of Oak timber grown in various places, says: "The more *sappy* timber is, the more it is subject to be infected with fungi and the dry-rot; thus all the timber brought from the *forests* of Germany, of which the Antwerp fleet has been built, is remarkably subject to the dry-rot; so is all the timber brought from the *forests* of America; whereas the timber of trees that grow in exposed situations, as on the sides of hills, and commons, and hedge-rows, being more compact, and less sappy, is less subject to this fatal disease. Soil and climate have also, no doubt, considerable influence on the nature of growing timber; the farther south Oak grows, the better the timber would seem to be; the Oak on the bold shores of the Adriatic is the best Oak known in Europe; and the Oak timber which is produced in the southern counties of England is preferred to the timber of the northern counties; that of Sussex being considered as the best. In a contract for tree-nails, drawn up more than a hundred years ago, it is stipulated that they should be made of 'good Sussex Oak, free from knots and slakes.'"

Much difference of opinion exists as to which species of British Oak produces the best timber. Early writers on the subject claim the superiority for *Q. robur*, or the "old English Oak," as they call it, on the ground that it is of more rapid growth, has a cleaner stem, and fewer knots, is more durable, and contains a larger proportion of heart-wood than the other species, *Q. sessiliflora*, or Durmast Oak. More recent authors, however, maintain that the true "old English Oak" is *Q. sessiliflora*, and account for the fact that it is now less common than the other on the supposition that our forefathers were well aware of the superiority of the former species, and applied it so extensively to all works requiring durability, that it has long become comparatively scarce. But a few years since, it was generally believed that the beautiful carved roof of Westminster Abbey was constructed of chestnut. Recent examination has, however, proved that it is composed entirely of Durmast Oak. This roof has stood for more than three hundred years. The foundation on which the stone piers of old London bridge were laid consisted of huge piles of timber, which when taken up were found to be perfectly sound, though they must have been driven upwards of six hundred years. The wood employed is from trees of the same species. Most of the timber found in old buildings which was formerly believed to be chestnut, is now known to be the wood of the Durmast Oak. In the year 1844, there was raised from the bottom of a lake at Davey Strand, between Dublin and Cavan, a huge canoe, which had been hollowed out of the trunk of a tree of the same kind. It measured no less than forty feet in length, the bottom being four feet three inches in diameter at one end, and about three feet at the other. On a fair computation, the circumference of this tree must have been at least twenty-one feet at the base, and fifteen feet at the height of forty feet from the ground. The antiquity of this relic is almost too great to be speculated on. Much of the wood-work in the old border-fortresses of Wales, and the doors of pews in ancient churches, are made from the same tree. The principal difference apparent to the eye between the timber of the two species is, that *Q. robur* is plentifully furnished with medullary rays, called by carpenters "silver-grain," of which the other species is almost entirely destitute, resembling in this respect the chestnut: from this similarity have probably sprung the numerous mistakes of the one wood for the other. On the whole, it would seem, that whatever good quality is found in either of the species, the other possesses in a somewhat greater or less degree, and there is little doubt that both will long continue to be applied indifferently to purposes where solidity, strength, and durability are required.

But the Oak begins to be valuable long before it has attained such a size as renders it fit for ship and house building. "The ground Oak, while young, is used for poles, cudgels, and walking-staffs, much come into mode of late, but to the waste of many a hopeful plant which might have proved good timber; and I the rather declaim against the custom, because I suspect they are such as are for the most part cut and stolen by idle persons, and brought up to London in great bundles, without the knowledge or leave of the owners, who would never have gleaned their copses for such trifling uses."*

According to Loudon, the proper age at which Oak copse should be cut down varies from "fifteen to thirty years; the rule being, that the principal stems of the plants, at one foot from the ground, should not be less than six inches in diameter. In favourable soils in the south and west of England, this size will be obtained in from twelve to fifteen years; but in the colder climate, and in the inferior soil of the Highlands of Scotland, from twenty-five to thirty years are required. The cutting over of copse is performed at the same season as that in which full-grown trees are felled, when in both cases the bark is an object as well as the timber." "The very stump of an Oak," says Evelyn, "especially that part which is dry and above ground, being well-grubbed, is

* Evelyn.

many times worth the pains and charge for sundry rare and hard works; and where timber is dear, I could name some who, abandoning this to workmen for their pains only, when they perceived the great advantage, repented of their bargain, and undertaking it themselves, were gainers above half. They made cups of the roots of Oaks heretofore; and such a curiosity, Athenæus tells us, was carved by Thericleus himself; and there is a way so to tinge Oak, after long burying and soaking in water, which gives it a wonderful politure, as that it has frequently been taken for a coarse ebony."*

The timber-merchant and the painter, if called to give an opinion on any particular Oak, would, in all probability, greatly differ. To the former a clean, straight, and regular stem would suggest calculations as to the number of cubic feet of timber it would be found to contain when the axe and square and saw should have done their work. A well-grown tree, therefore, in the vigour of its age, will be to him the perfection of all trees. The painter will, perhaps, stop and admire the stately growth of the same tree; he will notice the symmetry of its form, and watch the brilliant lights playing about its thick foliage; but he will feel no desire to transfer it to his canvass. There must be no perpendicular or parallel lines about the object of his choice; no semicircular evenly-shaped head; no arms of equal diameter springing from the main stem at the same angle, and extending to an equal distance all round. But shew him a veteran patriarch, whose gnarled trunk

* The "Black Oak," found in great quantities in many of the bogs in Ireland, might readily be mistaken for ebony.

is eaten out by the frosts of centuries, whose knotted limbs are fringed with ferns, and mottled with innumerable mosses and lichens; even if but a scanty foliage clings to branches which have been shattered again and again by the tempest; or if, instead of a leafy summit, it rears aloft a fantastic assemblage of hoary, sapless antlers;—and you will hear him exclaim, "I go no farther to-day; this is the tree for a picture!" And move he will not, until with his pencil he has produced the same image which the poet has conjured up with his pen.

"A huge Oak, dry and dead,
Still clad with reliques of its glories old,
Lifting to Heaven its aged, hoary head;
Whose foot on earth hath got but feeble hold,
And, half disbowelled, stands above the ground;
With wreathed roots, and naked arms,
And trunk all rotten and unsound."

SPENSER.

Gilpin (and few will be bold enough to differ from him in this respect) considers the Oak as the most picturesque of trees. He thus recommends to the artist a careful study of the various tints observable on its bark:—"I have often stood with admiration before an old forest Oak, examining the various tints which have enriched its furrowed stem. The genuine bark of an Oak is of an ash-colour, though it is difficult to distinguish any part of it from the mosses that overspread it; for no Oak, I suppose, was ever without a greater or less proportion of these picturesque appendages. The lower parts, about the roots, are often possessed by that green velvet moss which in a still greater degree commonly occupies the bole of the Beech; though the beauty and brilliancy of it lose much when in decay. As the trunk rises you see the brimstone colour taking possession in patches. Of this there are two principal kinds: a smooth sort, which spreads like a scarf over the bark; and a rougher sort, which hangs in little rich knots or fringes. I call it a brimstone hue, by way of general distinction; but it sometimes inclines to an olive, and sometimes to a light-green. We find also another species of moss of a dark-brown colour, inclining nearly to black; another of an ashy colour, and another of a dingy yellow. We may observe also touches of red, and sometimes, but rarely, a bright yellow, which is like a gleam of sunshine; and in many trees you will see one species growing upon another, the knotted brimstone-coloured fringe clinging to a lighter species, or the black softening into red. Strictly speaking, many of these excrescences, which I have mentioned under the general name of mosses, should have been distinguished by other names. All those, particularly, which cling close to the bark of trees, and have a leprous appearance, are classed, I believe, by botanists, under the name of lichens; others are called liver-worts. But all these excrescences, under whatever name distinguished, add a great richness to trees; and when they are blended harmoniously, as is generally the case, the rough and furrowed trunk of an old Oak, adorned with these pleasing appendages, is an object which will long detain the picturesque eye."

But with what a different eye would the timber-merchant look on these hollow trunks, and blasted antlers, and mottled lichens! He would see in them so much solid timber spoiled, so many knees and elbows rendered useless, and would count the cost of clearing the bark of so many superficial feet of nuisances. But, for this we must not quarrel with him ; nor, in our love of Nature, forget her useful subserviency to the arts of civilized life.

About the end of April the season for barking commences; and to this process Oaks both old and young are equally subjected; those of from twenty to thirty years' growth, however, being preferred. Oak bark is occasionally used in medicine, and is employed also as a dye, but is most valuable for the principle called tannin, which is indispensable in the manufacture of leather. Every part of the tree, indeed, abounds in astringent matter, and even the leaves and sawdust will tan leather, linen cloth, and netting or cordage, which is to be much exposed to the weather.

Melancholy though the sight is, when, resorting to some favourite woodland haunt, one encounters a dreary assemblage of naked wooden poles, instead of a grove of Oaks just budding into life; "yet the various appendages of woodcutting, piles of bark, and scattered boughs, and timber wains, are not unpleasing objects. The deep, hollow tone also of the woodman's axe, or of axes responsive to each other in different parts of the wood, are notes in full harmony with the scene, though their music is a knell. The fallen tree, also, lying with its white peeled branches on the ground, is not only beautiful in itself, but if it be not scattered in too great profusion, (for white is an unaccommodating hue,) it forms an agreeable contrast with the living trees. But when we see it deprived of its beautiful ramification, squared, and sawn in lengths, as it sometimes continues long to lie about the forest, it becomes an object of deformity; and we lament what it once was, without receiving any equivalent from its present state."*

A due supply of timber for the use of the Navy has long been a matter of consequence. In the Report of the Commissioners of land-revenue respecting timber, which was printed by order of the House of Commons in June 1812, it is stated that it requires three thousand loads of timber or two thousand well-grown Oak trees to build a seventy-four-gun ship. If we allow forty such trees to an acre, which is the highest number possible, even supposing the ground to be covered with trees all fit for naval purposes, the produce of fifty acres during a century will be required for a single seventy-four. For very many years, therefore, the Royal Forests, originally set apart for the amusement of the chase, have been jealously preserved as nurseries for timber. Of these there are a great number scattered throughout the various counties of England. The majority, however, exist only in name, having given way to the plough, to pasturage, to shipbuilding, or to the iron-foundry. Of the few which remain, the Forest of Dean, and the New Forest, are most worthy of mention. The first of these is in Gloucestershire, and has always been a place of note. It is of large extent, being not less than twenty miles in length, and half as many in breadth; and is separated from the rest of the county by the river Severn. The timber of this having been formerly in great request

* Gilpin's Forest Scenery.

for the purposes of ship-building, Evelyn tells us that he heard, "that in the great expedition of 1588, it was expressly enjoyned the Spanish commanders of that signal Armada, that if, when landed, they should not be able to subdue our nation, and make good our conquest, they should yet be sure not to leave a tree standing in the Forest of Dean. It was like the policy of the Philistines, when the poor Israelites went down to their enemies' smiths to sharpen every man his tools: for they said, 'lest the Hebrews make them swords or spears;' so these, 'lest the English build them ships and men-of-war.'" But though the Forest of Dean escaped this terrible denunciation, it was so effectually dismantled in the troublous times of Charles I., that on a survey made by order of Parliament in 1667, only two hundred trees were found standing. To repair these mischiefs, eleven thousand acres were immediately enclosed and planted; and it is from these that the supply of the dock-yards is now principally obtained, averaging about a thousand loads a-year. The riches of this Forest are not confined to the surface, there being large mines of iron and coal beneath nearly its whole extent.* Nearly fifteen thousand trees are annually delivered to the free miners and colliers for the carrying on of their works.

"The New Forest in Hampshire was originally made a forest by William I., in the year 1079, about thirteen years after the battle of Hastings,

* A well executed and beautiful model of the Forest of Dean may be seen in that interesting, but as yet little known exhibition, the Museum of Economic Geology, in Craig's-court, Scotland-yard, near Charing-cross, London.

and is, indeed, the only forest in England whose origin can be traced. It took the denomination of New Forest, from its being an addition to the many forests which the Crown already possessed, and which had formerly been appropriated in feudal times."* Its soil being peculiarly adapted to the growth of Oak, it has been for centuries considered one of the great magazines for the Navy. Its resources being considered inexhaustible, enormous quantities of timber were taken away, but no care taken to ensure a future supply by planting. Thus, in Charles the Second's time, the nation, being on the eve of a war with the Dutch, began to find themselves at a loss to supply the urgent demand for building materials. In 1664, Evelyn, at the request of the Royal Society, wrote his *Sylva*, in which he enforces the necessity of some extensive system of planting forest-trees. The result was, that in 1669 the first royal mandate was issued for enclosing and planting a portion of the New Forest as a nursery for young Oaks; "and," says Hunter, "the spirit for planting increased to a high degree; and there is reason to believe that many of our ships, which, in the last war, gave laws to the whole world, were constructed from Oaks planted at that time." In the reign of King William the Third, an act was passed empowering certain commissioners to enclose two thousand acres in the New Forest for the growth of timber, and two hundred more every year for the space of twenty years afterwards. Equally active measures have been adopted in the present century, so that the New Forest bids fair to outlive the injudicious attacks of do-

* Gilpin.

mestic friends, as the Forest of Dean did the fury of foreign enemies.

In the New Forest formerly stood an Oak, from which the arrow glanced that killed William Rufus. Like the Cadenham Oak before mentioned, it came into leaf at Christmas, and, although it has long since disappeared, it is remarkable that a young tree, near the spot where it stood, is subject to the same peculiarity. The veteran tree was paled round by the command of Charles II., in order to preserve it, and it stood until at least the commencement of the 18th century; for in the year 1745, to mark the spot, a triangular stone was erected on it by Lord Delaware, on the three sides of which were the following inscriptions:—

"Here stood the Oak tree on which an arrow, shot by Sir Walter Tyrrell at a stag, glanced and struck King William II., surnamed Rufus, on the breast, of which stroke he instantly died, on the 2nd of August, 1100." "King William II., being thus slain, was laid in a cart belonging to one Purkess, and drawn from hence to Winchester, and buried in the cathedral church of that city." "That the spot where an event so memorable happened might not hereafter be unknown, this stone was set up by John Lord Delaware, who has seen the tree growing in this place." * This inscription is now nearly effaced.

There is a particular interest connected with trees of great antiquity, which attaches itself to nothing else. A flourishing Oak in the vigour of its age, furnished with a well-proportioned, tapering trunk, and with symmetrically-arranged

* Gilpin. See Frontispiece.

branches, and flinging its chequered shade far and near over the verdant sward, is a beautiful object, and irresistibly draws the attention to itself. But it does not carry the mind of the spectator back to past events, — it does not talk with us about bygone ages, and scenes at which no man now living was present. And, if we think of its future fate, there is so much of uncertainty about that, so much of doubt as to the length of time for which it is destined to retain its position; whether it will be laid low by the tempest, or by the woodman's axe, and if the latter, to what purposes it may be applied, that the mind can select nothing sufficiently definite to engage itself upon. The tan-yard, the saw-pit, and the baker's oven, are decidedly not subjects to dwell upon; and these, in fact, are the only passages in its history which can be predicted with certainty. But the case is very different with the uncouth monster on whom the destroyer has done all but his utmost. Though but a hollow shell, blasted above, and worm-eaten below, and indebted for its scanty verdure more to ferns and moss than to the feeble relics of life which yet remain in it; it is a monument of the past more eloquent than buildings the most time-hallowed; or, save one, than books of the most remote antiquity. It is *now* a *living* tree, and it was the same thirty generations back. Yes! a thousand years ago it was a stately tree: — when the present dynasty commenced it was older than the oldest men then alive, and it has lived through all the stirring events which have taken place from that time to this, connecting the names of Scott and Wordsworth with those of Newton and Milton, and Shakspeare, and these with Spenser and

Chaucer; and having sprung from an acorn borne by a tree which perhaps flourished when our holy religion was preached in Palestine by the Saviour, whose coming was to banish from the earth all those barbarous rites which were then being enacted beneath the shade of its branches.

This is not all purely imaginary; the evidence in favour of an antiquity nearly as great as that assumed being as conclusive as the necessarily imperfect records will admit.

"Close by the gate of the water-walk at Magdalen College in Oxford, grew an Oak, which perhaps stood there a sapling when Alfred the Great founded the University. This period only includes a space of nine hundred years, which is no great age for an Oak. It is a difficult matter indeed to ascertain the age of a tree. The age of a castle or abbey is the object of history; even a common house is recorded by the families that built it. All these objects arrive at maturity in their youth, if I may so speak. But the tree, gradually completing its growth, is not worth recording in the early part of its existence. It is then only a common tree; and afterwards, when it becomes remarkable for its age, all memory of its youth is lost. This tree, however, can almost produce historical evidence for the age assigned to it. About five hundred years after the time of Alfred, William of Wainfleet, Dr. Stukely tells us, expressly ordered his college to be founded near the Great Oak; and an Oak could not, I think, be less than five hundred years of age to merit that title, together with the honour of fixing the site of a college. When the magnificence of Cardinal Wolsey erected that handsome tower which is so ornamental to the whole building, this tree might probably be in the meridian of its glory, or rather, perhaps, it had attained a green old age. But it must have been manifestly in its decline at that memorable era when the tyranny of James gave the fellows of Magdalen so noble an opportunity of withstanding bigotry and superstition. It was afterwards much injured in Charles the Second's time, when the present walks were laid out. Its roots were disturbed, and from that period it declined fast, and became reduced by degrees to little more than a mere trunk. The oldest members of the University can scarce recollect it in better plight. But the faithful records of history have handed down its ancient dimensions. Through a space of sixteen yards on every side from its trunk, it once flung its boughs, and under its magnificent pavilion could have sheltered with ease three thousand men, though in its decayed state it could for many years do little more than shelter some luckless individual whom the driving shower had overtaken in his evening walk. In the summer of the year 1788, this magnificent ruin fell to the ground, alarming the College with its rushing sound. It then appeared how precariously it had stood for many years. Its grand tap-root was decayed, and it had hold of the earth only by two or three roots, of which none was more than a couple of inches in diameter. From a part of its ruins a chair has been made for the President of the College, which will long continue its memory." *

Among several celebrated Oaks in Windsor Forest, Loudon mentions one, called the King

* Gilpin.

Oak, which "is said to have been a favourite tree of William the Conqueror, who made this a royal forest, and enacted laws for its preservation. This Oak, which stands near the enclosure of Cranbourn, is twenty-six feet in circumference at three feet from the ground. It is supposed to be the largest and oldest Oak in Windsor Forest, being above a thousand years old. It is quite hollow: the space within is from seven to eight feet in diameter, and the entrance is about four feet and a half high, and two feet wide. 'We lunched in it,' says Professor Burnet, 'Sept. 2nd, 1829: it would accommodate at least twenty persons with standing-room: and ten or twelve might sit down comfortably to dinner.'"

The Winfarthing Oak, in Norfolk, claims an origin yet higher, and is still standing. A writer in the Gardener's Magazine gives the following account of this remarkable tree. "Of its age I regret to be unable to give any correct data. It is said to have been called the 'Old Oak' at the time of William the Conqueror, but upon what authority I could never learn. Nevertheless, the thing is not impossible, if the speculations of certain writers on the age of trees be at all correct. Mr. South, in one of his letters to the Bath Society (vol. x.), calculates that an Oak tree forty-seven feet in circumference cannot be less than fifteen hundred years old; and Mr. Marsham calculated the Bentley Oak, from its girting thirty-four feet, to be of the same age.* Now, an inscription on a

* Lengthy and somewhat abstruse calculations have been made for ascertaining the age of a tree from the diameter of its trunk. They are not, however, much to be depended on, the rate of growth often varying to a very great degree even in trees planted at the same time

brass plate affixed to the Winfarthing Oak gives us the following as its dimensions: 'This Oak, in circumference, at the extremities of the roots, is seventy feet; in the middle, forty feet, 1820.' Now, I see no reason, if the size of the rind is to be any criterion of age, why the Winfarthing should not, at least, equal the Bentley Oak; and if so, it would be upwards of seven hundred years old at the Conquest; an age which might very well justify its then title of the 'Old Oak.' It is now a mere shell, a mighty ruin, bleached to a snowy white, but it is magnificent in its decay. The only mark of vitality it exhibits is on the south side, where a narrow strip of bark sends forth a few branches, which even now occasionally produce acorns. It is said to be very much altered of late; but I own I did not think so when I saw it about a month ago (May 1836); and my acquaintance with the veteran is of more than forty years' standing: an important portion of *my* life, but a mere span of its own."

"In a glade of Hainhault Forest, in Essex, about a mile from Barkingside, stands an Oak, which has been known through many centuries by the name of Fairlop. The tradition of the country traces it half-way up the Christian era. It is still a noble tree, though it has now suffered greatly from the depredations of time. About a yard from the ground, where its rough fluted stem is thirty-six feet in circumference, it divides into

and in the same soil. The increase of diameter varies yet again when the trunk has become even partially hollow; for in this case the pressure arising from the new wood which is deposited under the bark acts in an inward as well as an outward direction; consequently the annual increase of diameter is retarded, though not in a known ratio.

eleven vast arms, yet not in the horizontal manner of an Oak, but in that of a Beech. Beneath its shade, which overspreads an area of three hundred feet in circuit, an annual fair has long been held on the 2nd of July, and no booth is suffered to be erected beyond the extent of its boughs. But as their extremities are now become sapless, and age is yearly curtailing their length, the liberties of the fair seem to be in a very desponding condition. The honour, however, is great. But honours are often accompanied by inconveniences, and Fairlop has suffered from its honourable distinctions. In the feasting that attends a fair, fires are often necessary, and no places seemed so proper to make them in as the hollow cavities formed by the heaving roots of the tree. This practice has brought a speedier decay on Fairlop than it might otherwise have suffered."*

Phillips, in his *Companion to the Orchard*, adds the following notice of the same tree. "This venerable Oak was cut down previous to the fair in 1820. The founder of this fair was a Mr. Daniel Day, commonly called the Good Day, who was born in the parish of St. Mary Overy, in 1682; his father was an opulent brewer, but Mr. Day followed the business of a block and pump maker in Wapping, and possessing a small estate in Essex, at no great distance from this remarkable tree, he used, on the first Friday in July, annually to repair thither, having given his accustomed invitation to a party of his neighbours to accompany him, for the purpose of dining under the shade of its branches and leaves on beans and bacon. This benevolent as well as humorous

* Gilpin.

man never failed to pay his annual visit to the public bean-feast, and as regularly provided several sacks of beans, with a proportionate quantity of bacon, which he distributed from the trunk of the tree to the persons there assembled. A few years before the decease of Mr. Day, (in 1767,) his favourite Oak lost a large limb, out of which he procured a coffin to be made for his own interment. We have been informed that the following circumstance gave rise to the name of Fairlop, bestowed upon this celebrated Oak. Some of Mr. Day's friends having promised that he should be buried in a coffin made from that tree, lopped off one of the branches, for which trespass an action was brought against the party, fortunately for whom some flaw was found in the pleadings, and the plaintiff was nonsuited. It was, however, proved that the fact committed was not injurious to the tree, but a *fair lop*. As lately as 1794, this venerable Oak, in the meridian of the day, shadowed an acre of ground, although then greatly decayed."

Some years before its fall Mr. Forsyth's composition was applied to its decayed branches to preserve it from future injury; when a board was affixed to one of its limbs, bearing the following inscription: — "All good foresters are requested not to hurt this old tree, a plaster having been lately applied to his wounds." In the year 1805 its trunk took fire, in consequence of the carelessness of a party of cricketers, who had spent the day in its vicinity, and had left a fire burning too near it. The fire was discovered the same evening; and although a number of persons did their utmost to extinguish the flames, it continued burning till morning. This untoward accident so weakened it, that, as Professor Burnet informs us, "the high winds of February, 1820, stretched this forest patriarch on the ground, after having endured the storms of perhaps one thousand winters. Its remains were purchased by a builder; and from a portion thereof the pulpit and reading-desk in the new church, St. Pancras, were constructed: they are beautiful specimens of British Oak, and will long preserve the recollection of this memorable tree." The largest Oak on record grew in Dorsetshire. It was called Damory's Oak, and was used as an ale-house. It was sixty-eight feet in circumference, and the room formed in it was sixteen feet in length. This tree was blown down in 1703.

The celebrated Chapel-Oak of Allonville in the Pays de Caux, in France, which is still standing, measures at its base thirty-five feet in circumference, and at six feet above the level of the ground it is twenty-six feet in girth. It is hollow, and the interior is fitted up as a chapel. This transformation was effected in 1696. The computed age of the tree is between eight and nine centuries.

The largest Oak now existing of which I can find any account is the Cowthorpe Oak, near Wetherby in Yorkshire. Of this tree a spirited engraving is given in Hunter's Evelyn's Sylva,* together with the following description. "Within three feet of the surface it measures sixteen yards in circumference, and close by the ground, twenty-six yards. Its height is about eighty feet, and its principal limb extends sixteen yards from

* Vol. ii. p. 197.

the bole. Throughout the whole tree the foliage is extremely thin, so that the anatomy of the antient branches may be distinctly seen in the height of summer. This venerable tree must once have been the pride of the forest, but now

> ——— the gray moss mars his rime,
> His bare boughs are beaten with stormes,
> His top is bald and wasted with wormes,
> His honour decayed, his branches sere."
>
> SPENSER.

The drawing of this tree was made in 1776, and the description published ten years later. The following account was sent to Loudon in 1829:—"On a stranger's first observing the tree he is struck with the majestic appearance of its ruined and riven-looking dead branches, which in all directions appear above the luxuriant foliage of the lateral and lower arms of the tree. In 1722 one of the side-branches was blown down in a violent gale of wind, and, on being accurately measured, was found to contain upwards of five tons of wood. The largest of the living branches at present extends about forty-eight feet from the trunk; and its circumference, at about one yard from the giant bole, is eight feet six inches. Three of the living branches are propped by substantial poles, resting upon stone pedestals. The diameter in the hollow part at the bottom is nine feet ten inches: the greatest height of the dead branches is about fifty-six feet. It is evidently of very great antiquity, as all tradition represents it as a very old tree." "The circle occupied by the Cowthorpe Oak," says Professor Burnet, "where the bottom of its trunk meets the earth, exceeds the ground-plot of that majestic column of which an Oak is confessed to have been the prototype, viz., Smeaton's Eddystone light-house. Sections of the trunk of the one would, at several heights, nearly correspond with sections of the curved and cylindrical portions of the shaft of the other. Arthur's round table would form an entire roof, or projecting capital, for the lighthouse: indeed, upon this table might be built a round church as large as that of St. Lawrence, in the Isle of Wight, and space to spare; so that, if the extent of the sap-wood be added, or the ground-plot of the Cowthorpe Oak be substituted for Arthur's table, there would be plenty of room, not only to build such a parish church, but to allow space for a small cemetery beside it. Indeed," continued Burnet, "I would merely observe that St. Bartholomew's, in the hamlet of Kingsland, between London and Hackney, which, beside the ordinary furniture of a place of religious worship, has pews and seats for one hundred and twenty persons, is nearly nine feet less in width, and only seventeen inches more in length, than the ground-plot of the Cowthorpe Oak. In fact, the tree occupies upwards of thirty square feet more than does the chapel."

The following affecting story is told by White: —"In the centre of Josel's wood there formerly stood an Oak, which, though stately and tall on the whole, bulged out into a large excrescence about the middle of the stem. On this a pair of ravens had fixed their residence for such a series of years that the Oak was distinguished by the title of the Raven Tree. Many were the attempts of the neighbouring youths to get at this eyry; the difficulty whetted their inclinations, and each was ambitious of surmounting the arduous task. But when they arrived at the swelling, it jutted out so in their way, and was so far beyond their grasp, that the most daring lads were awed, and acknowledged the undertaking to be too hazardous. So the ravens built on, nest upon nest, in perfect security, till the fatal day arrived in which the wood was to be levelled. It was in the month of February, when those birds usually sit. The saw was applied to the butt, the wedges were inserted into the opening. The woods echoed to the heavy blows of the beetle, or mallet; the tree nodded to its fall, but still the dam sat on. At last, when it gave way, the bird was flung from her nest; and, though parental affection deserved a better fate, was whipped down by the twigs, which brought her dead to the ground."*

The Oaks most remarkable for their horizontal expansion, are, according to Loudon, the following. "The Three-shire Oak, near Worksop, was so situated, that it covered part of the three counties of York, Nottingham, and Derby, and dripped over seven hundred and seventy-seven square yards. An Oak between Newnham Courtney and Clifton shaded a circumference of five hundred and sixty yards of ground, under which two thousand four hundred and twenty men might have commodiously taken shelter. The immense Spread Oak in Worksop Park, near the white gate, gave an extent, between the ends of its opposite branches, of an hundred and eighty feet. It dripped over an area of nearly three thousand square yards, which is above half an acre, and would have

* White's Natural History of Selborne.

afforded shelter to a regiment of nearly a thousand horse. The Oakley Oak, now growing on an estate of the Duke of Bedford, has a head of an hundred and ten feet in diameter. The Oak called *Robur Britannicum*, in the Park at Rycote, is said to have been extensive enough to cover five thousand men; and at Ellerslie, in Renfrewshire, the native village of the hero Wallace, there is still standing 'the large Oak tree,' among the branches of which it is said that he an three hundred of his men hid themselves from the English."

In addition to this last, there are many old Oaks which possess a legendary interest. Gilpin tells us that, "in Torwood, in the county of Stirling, upon a little knoll, stands at this time the ruins of an Oak, which is supposed to be the largest tree that ever grew in Scotland. The trunk of it is now wholly decayed and hollow, but it is evident from what remains, that its diameter could not have been less than eleven or twelve feet. What its age may be, is matter only of conjecture; but, from some circumstances, it is probably a tree of great antiquity. The little knoll it stands on is surrounded by a swamp, over which a causeway leads to the tree, or rather to the circle which seems to have been round it. The vestiges of this circle, as well as of the causeway, bear a plain resemblance to those works which are commonly attributed to the Druids; so that it is probable this tree was a scene of worship belonging to those heathen priests. But the credit of it does not depend on the dubious vestiges of Druid antiquity. In a later scene of greater importance (if tradition ever be the vehicle of truth), it bore a great share. When that illustrious hero, William

Wallace, roused the spirit of the Scotch nation to oppose the tyranny of Edward, he often chose the solitude of Torwood as a place of rendezvous for his army. Here he concealed his numbers and his designs, sallying out suddenly on the enemy's garrisons, and retreating as suddenly when he feared to be overpowered. While his army lay in these woods, the Oak which we are now commemorating, was commonly his head-quarters. Here the hero generally slept; its hollow trunk being capacious enough to afford shelter, not only to himself, but to several of his officers. This tree has ever since been known by the name of Wallace Tree, by which name it may be easily found in Torwood to this day."

Another interesting relic is the Parliament Oak, which grows in Clipstone Park, and takes its name from the fact of a Parliament having been held under it, by Edward I., in 1290.

"Queen Elizabeth's Oak, at Huntingfield, in Suffolk, measures thirty-four feet in girth at five feet from the ground. Queen Elizabeth is said to have been entertained at the old mansion by Lord Hunsdon, and to have enjoyed the pleasures of the chase in rural majesty. The great hall was built round six straight massy Oaks, which originally upheld the roof, as they grew; and upon these the foresters and yeomen of the guard used to hang their nets, cross-bows, hunting-poles, and other implements of the chase. Elizabeth is said to have been much pleased with the retirement of this park, filled with tall and massy timber-trees, but particularly with the Oak, which ever afterwards bore the appellation of the Queen's Oak. It stood about two bow-shots from the old romantic hall; and tradition records, that Elizabeth shot a buck, with her own royal hand, from this tree."*

Evelyn, who wrote his Sylva in the reign of Charles II., thus dedicated the Fourth Edition to that Monarch. "To you then, Royal Sir, does this Fourth Edition continue its humble addresses, since you are our *Nemorensis Rex:* as having once had your temple, and court too, under that sacred Oak which you consecrated with your presence, and we celebrate, with just acknowledgement to God, for your preservation."

The tree here alluded to, called the "Royal Oak," formerly stood at Boscobel, in Shropshire, but was destroyed soon after it attained its notoriety by the ill-judged curiosity of the Royalists. For the same author, speaking of an Oak which put forth its buds about Christmas, says:—King James went to visit it, and "caused benches to be placed about it; which giving it reputation, the people never left hacking of the boughs and bark till they killed the tree: as I am told they have served that famous Oak near White-Lady's, which hid and protected our late Monarch from being discovered and taken by the rebel soldiers who were sent to find him, after his almost miraculous escape at the battle of Worcester." In the course of this spoliation a huge bulk of timber, consisting of many loads, was carried away in handfuls. Several saplings were raised in different parts of the country from its acorns, one of which grew near St. James's Palace, where Marlborough House now stands; and there was another in the

* Lauder's Gilpin.

Botanic Garden, Chelsea. The former has been long since felled; and of the latter even the recollection seems now almost lost.

Through the kindness of the Rev. J. Dale, Curate of Donington, the parish in which the Boscobel Oak stands, I am enabled to lay before my readers a full and authentic account of a tree, which, from its connexion with one of the most important events in English History, will always be remembered with interest.

On a single printed leaf which is pasted in at the end of one of the Parish Registers of Donington, is the following note, in the handwriting of the late Rector, Dr. Woodhouse. "*Extracts from the Philosophical Transactions, vol. 5, part 2nd, chap. 3, written by the Rev. George Plaxton, Rector of Donington (and Kinnardsey) from* 1690 *to* 1703." Then follows the type. "The Royal Oak was a fair spreading tree; the boughs of it all lined and covered with ivy. Here, in the thick of these boughs, the King sat in the day-time, with Colonel Carlos, and in the night lodged in Boscobel House; so that they are strangely mistaken who judged it an old hollow oak, whereas it was a gay and flourishing tree surrounded with a great many more, and, as I remember in Mr. Evelyn's History of Medals, you have one of King James I. or Charles I. where there is a fine spread Oak with this epigraph, 'Seris nepotibus umbra,' which I leave to your thoughts. * * * The poor remains of the Royal Oak are now fenced in by a handsome brick wall, at the charge of Basil Fitzherbert, Esq., with this inscription over the gate, upon a blue stone in letters of gold.

Felicissimam arborem quam in asylum
potentissimi Regis Caroli Secundi Deus Opt. Max.
per quem Reges regnant hic crescere
voluit tam in perpetuam rei tantæ
memoriam quam in Specimen firmæ
in reges fidei muro cinctam
posteris commendant Bazillius
et Jana Fitzherbert.
Quercus amica Jovi.*

"'Twas put up about twenty or thirty years ago; but the place deserved a better memorial. I have writ it in such lines as they have cut it, and as the letters now stand; a few years will ruine both the wall and the inscription."

"The emblematical medal my good friend alludes to is the XLVII. in Mr. Evelyn's Numismata, which King Charles caused to be stamped in honour of the installation of his son; whereupon is the Royal Oak under a Prince's coronet, overspreading subnascent trees and young suckers."

In the year 1812 or thereabouts, and before he was aware of Mr. Plaxton's notice, Mr. Dale discovered portions of the above inscription "on a blue stone, in letters of gold," among the long and neglected grass on the Mount in Boscobel Garden. After spending some time in arranging the fragments, he communicated the discovery to the occupants of the house, who appear to have taken little interest in the relic. The house and grounds have passed into other hands, and the fragments of the stone in all probability lie buried beneath the present garden walks, which were laid out

* *Translation.*—This most highly favoured tree, planted by the God through whom kings reign to afford shelter to his Majesty King Charles the Second, was enclosed with a wall by Basil and Jane Fitzherbert, as well to preserve to posterity a memorial of the auspicious event as to be a token of their own steadfast loyalty.

by the present proprietor after the pattern of those which existed in the time of Charles II. Of the tree itself very few, and these imperfect, records remain. Old Plaxton speaks of it as "a fair spreading tree, the boughs of it all lined and covered with ivy," and that in the thick of it the King and Carlos sat. This agrees well with the description of it which the King himself gives in his narrative, "A great Oak that had been lopped some three or four years before, and, being grown out again very bushy and thick, could not be seen through; and here we staid all the day." This would be an excellent hiding-place; for, says Mr. Dale, "I have frequently observed that an old Pollard Oak, standing on a bank and overhanging the road between the Churches of Albrighton and Donington, about one hundred yards from each, would afford a secure retreat for two or three persons, from the observation of all passers by."

It will be seen by the extract from Evelyn's Sylva, that in 1662 it had ceased to be a living monument of the event to which it owes its celebrity. Not many years after, its poor remains were fenced in by a "handsome brick wall;" but all in vain. Every vestige of the original tree has disappeared from the spot more than a century ago. Mr. Dale thinks, from inquiries made on the spot from persons whose age, if they were now alive, would exceed a hundred, that the last remnants were taken away about the year 1734.

The handsome brick wall above alluded to stood until the year 1817, having been repaired in 1787 by Basil and Eliza Fitzherbert, who also attached a new inscription. Mr. Dale has been unable to discover any written account of the second tree thus enclosed. By general tradition, however, it sprung from an acorn of the Royal Oak, and this is credible enough; for whoever took the pains to rear young trees for St. James's Park and the Chelsea Gardens, doubtless did all in his power to perpetuate the race on the spot where the event took place. From the inscription of 1787, it would seem that Basil and Eliza Fitzherbert believed the tree then standing to have been the identical one in which the Sovereign took shelter. But, although they were mistaken in this respect, it must have attained a considerable size, or they could not have fallen into such an error. From this and other circumstances it appears tolerably certain that the tree now standing is the immediate descendant of the Royal Oak, and that it was planted about the time of the Restoration in 1660, as nearly in the same site as the remains of the old tree would allow, some of the old people alluded to above recollecting that it did not stand in the centre of the old enclosure.

The present Royal Oak, however, is now rapidly following its predecessor to decay. No casual observer would, from its appearance, suspect that, only fifty years ago, its branches spread over a spacious circumference, far beyond the wall, reaching to within a few feet of the ground, and so umbrageous that about that time a party of roguish rustics concealed themselves under its shade whilst on the watch for an outlying deer, as it came trotting up the green sward to browse on the tillage. Soon after that period, however, it suffered severely from the tempests, especially on one occasion about forty years ago. The branches, before they had lost their leaves, were loaded with snow, which became partially frozen, so that, when the wind got up, there was such a terrific crashing and mangling of its limbs that not less than a waggon-load and a half were carried to the wood-pile, and it appears never to have recovered from the effects of this unseasonable snow-storm; though it was observed at the time that other Oaks in the neighbourhood suffered comparatively little damage. Mr. Dale thinks, with reason, that having been cooped up for more than a hundred years within high walls, its timber did not acquire the hardness and toughness which a free exposure to the air would have given it. In 1817 the wall was taken down and iron palisades erected in its stead; but the remedy was applied too late. On its exposure several holes in the trunk were discovered, which, as well as others which followed the tearing off of its arms, were covered with lead, but without beneficial result. The over-nursed giant, which might under happier circumstances have now been in the pride of its strength, assumes yearly a more faded face and form; its leafy branches, in place of their former graceful sweep towards the ground, are contracting in circumference, so that now the lowest of them is not less than twenty feet from the earth. For the last thirty years it has been a shy bearer, not bringing acorns to perfection oftener than once in eight or ten years. About ten years ago Mr. Dale gathered a few, expecting they would be the last crop of ripe fruit. The seasons in the interval between that time and 1844 produced scarcely a handful of acorns each, none of which vegetated. The unusual heat and dryness of that summer, which, it is said, produced generally a larger crop of acorns than is ever remembered, revived its drooping powers, and it bore two or three pecks. "There is not any danger," Mr. Dale says, "of the race of the present Royal Oak becoming extinct, it having already double as many authenticated descendants as King Priam had I was myself the tutor of some scores of them in the year 1834-5, and have got several of them good appointments at the seats of Lord Dungannon, in Denbighshire, Sir Astley Cooper, in Hertfordshire, &c. More than thirty young plants, the produce of 1834, were this year planted upon the glebe at Albrighton, and a plentiful crop has been reared from the acorns of 1844. The largest known descendant of the present Royal Oak was planted* by the late

* The inscription attached to the present tree by the proprietor of the Boscobel estate, Miss Evans, is as follows:

Felicissimam arborem,
quam in asylum Potentissimi Regis Caroli II.
Deus Optimus Maximus per quem reges regnant
hic crescere voluit,
tam in perpetuam rei tantæ memoriam,
quam in specimen firmæ in reges fidei,
muro cinctam posteris commendârunt
Basilius et Jana Fitzherbert:
quod pietatis monimentum vetustate collapsum
paternarum virtutum hæredes
et avitæ in principes fidei æmulatores
in integrum restituerunt
Basilius et Eliza Fitzherbert,
iiii Cal. Junii, A.S. MDCCLXXVII.
Qua ex arbore hanc arborem, uti fertur, ortam
Ferreis his, quæ hodie sunt, repagulis
circummunivit
ejusdem hujusce agri possessor,
eodemque erga reges animo prædita,
Francesca Evans,
A.D. MDCCCXVII.

The tablet, bearing this inscription, was set up on the 25th May 1845.

Rector, Dr. Woodhouse, Dean of Lichfield, about seventy years ago. It stands one hundred and thirty-two yards from the N. E. angle of the Chancel of the Church at Donington, and measures three feet nine inches in diameter at four feet from the ground."

Some notion of the value of a well-grown Oak in its prime may be formed from the following account of the felling, in the year 1758, of a tree in Langley Wood, on the borders of the New Forest, and of another in Monmouthshire. The former of these, Mr. South tells us, "stood singly in the Wood, and extended its massive branches near forty feet each way. Its head was all knees and crooks, aptly suited to naval purposes; its bole or shaft was short, not exceeding twenty feet in length; but it was full six feet in diameter at the top, and perfectly sound. It was felled in an unusual manner for the preservation of its crooks, which were cut off one by one whilst the tree was standing, and lowered by tackles, to prevent their breaking. The two largest arms were sawed off at such distances from the bole as to make first-rate knees; scaffolds were then erected, and two pit-saws being braced together, the body was first cut across, half through, at the bottom, and then sawed down the middle, perpendicularly, between the two stumps of arms that had been left, at the end of one of which stood a perpendicular bough, bigger than most timber-trees. To prevent this being injured, a bed was made of some hundreds of faggots to catch it when it fell. This half was so weighty that it crushed a new timber-carriage all to pieces the instant it was lodged upon it; and, none in the country being found strong enough, the King's carriage was sent purposely from Portsmouth to convey it to the Dock-yard. This tree was sold in the first place for 40*l.*; was bought of that purchaser by a timber-merchant for 100*l.*, who is supposed to have cleared 100*l.* more; which he might very well do, for the contents amounted to thirty-two loads of hewed timber, which, at half-a-crown a foot—no unusual price for naval crooks—amounts to 200*l.* precisely, besides faggots, &c. sufficient to defray the expenses. The breadth of the tree across, near the ground, where it was cut, was twelve feet, and it had above three hundred rings of annual growth."

"The Gelonos Oak, which was cut down in 1810, grew about four miles from Newport, in Monmouthshire. The main trunk was ten feet long, and produced four hundred and fifty cubic feet of timber; one limb, three hundred and fifty-five feet; one ditto, four hundred and seventy-two feet; one ditto, one hundred and thirteen feet; and six other limbs of inferior size averaged ninety-three feet each, making a total of two thousand four hundred and twenty-six feet of convertible timber. The bark was estimated at six tons; but, as some of the very heavy body-bark was stolen out of the barge at Newport, the exact weight is not known. Five men were twenty days stripping and cutting down this tree; and two sawyers were five months converting it, Sundays excepted. The main trunk was nine and a half feet in diameter; and, in sawing it through, a stone was discovered six feet from the ground, above a yard in the body of the tree, through which the saw cut. The stone was about six inches in diameter, and was completely shut in;

but around it there was not the least symptom of decay. The rings in the butt were carefully counted, and amounted to upwards of four hundred in number; a convincing proof that this tree was in an improving state for upwards of four hundred years; and, as the ends of some of its branches were decayed and had dropped off, it is presumed that it had stood a great number of years after it had attained maturity. The bark of this tree, Burnet says, was sold by the merchant for the scarcely credible sum of 200*l.* This Oak was purchased for 100 guineas, under the apprehension of its being unsound; but Burnet tells us that it was resold while standing for 405*l.*, and that the cost of converting it was 82*l.*, amounting altogether to 487*l.*; it was subsequently resold for 675*l.* There were at least four hundred rings or traces of annular growth within its mighty trunk."*

There yet remains one Oak to be mentioned, which, if not so celebrated as any of the foregoing for size and expansion, is yet well worthy of note, as bearing testimony to the piety of our forefathers. This is the Gospel Oak, a boundary Oak which divides the parishes of Stoneleigh and Baginton. The name is common to many other trees, and is derived from the custom of perambulating the bounds of a parish on Rogation days by the inhabitants, in order that the localities might be impressed on the memories of the young. The Minister, accompanied by the Churchwardens, took the lead, and stopped at remarkable spots and trees to recite passages from the Gospel, the

* The account of this tree is furnished by Loudon, (Arboretum Britannicum, p. 1777,) being compiled from various sources.

103rd and 104th Psalms, with the litany and suffrages, and Homily of Thanksgiving.

"There were in my remembrance," says Evelyn, "certain prayers, litanies, and collects, solemnly used by the parish minister in the field, at the limits of their perambulations on the Rogation days, from an ancient and laudable custom of above one thousand years, introduced by Avitus, the pious Bishop of Vienna, in a great dearth, unseasonable weather, and other calamities, (however in tract of time abused by many gross superstitions and insignificant rites, in imitation of the Pagan Robigalia,) upon which days, about the Ascension and beginning of Spring especially, prayers were made, as well deprecatory of epidemical evils, amongst which blasts and smut of corn were none of the least, as supplicatory for propitious seasons and blessings on the fruits of the earth. Whether there was any peculiar office, besides those of Ember-weeks, appointed, I do not know; but the pious and learned Bishop of Winchester (Andrews) has, in his Devotions, left us a prayer so apposite and comprehensive for these emergencies, that I cannot forbear the recital.

"Remember, O Lord, to renew the year with thy goodness, and the season with a promising temper; for the eyes of all wait upon Thee, O Lord; Thou givest them meat; Thou openest thy hand, and fillest all things living with thy bounty. Vouchsafe, therefore, O Lord, the blessings of the heavens, and the dews from above: the blessings of the springs, and the deep from beneath: the returns of the sun, the conjunctions of the moon: the benefit of the rising mountains, and the lasting hills: the fulness of the earth, and all that breed therein:

A fruitful season,	Just laws,
Temperate air,	Righteous judgments,
Plenty of corn,	Loyal obedience,
Abundance of fruits,	Due execution of justice,
Health of body, and	Sufficient store for life,
Peaceable times;	Happy births,
Good and wise government,	Good and fair plenty,
Prudent counsels,	Breeding and institution of children,

that our sons may grow up as the young plants, and our daughters may be as the polished corners of the temple: that our garners may be full and plenteous with all manner of store: that our sheep may bring forth thousands: that our oxen may be strong to labour: that there be no decay: no leading into captivity: no complaining in our streets: but that every man may sit under his own vine, and under his own fig-tree, in thankfulness to Thee: sobriety, and charity to his neighbour: and in whatsoever state Thou wilt have him, therewith to be contented: and this for Jesus Christ his sake, to whom be glory for ever. Amen."

THE ILEX, EVERGREEN OAK, OR HOLM-OAK.

Quercus Ilex.

Natural order—Amentaceæ.

Class—Monœcia. *Order*—Polyandria.

The Ilex is not a native of Great Britain; nor, although it flourishes and becomes a large tree in congenial situations, is it likely that it will ever become so far naturalized as to propagate itself to any extent. Nevertheless, as an ornament to the landscape it is a great acquisition, affording in summer, with its sombre foliage, a pleasing contrast to the brighter tints of every other tree in the neighbourhood, and no less valuable when the deciduous trees have thrown off their perishable garniture, and wisely prepared themselves to encounter the storms of winter by clearing themselves of what would oppose their boisterous progress. The Ilex, too, will stand the sea-breeze uninjured, and thrives better than most other evergreens in the vicinity of cities where it is exposed to the action of coal-smoke. For all these reasons, therefore, now that more attention is paid to the subject of planting than ever was before, man will in all probability do for it what Nature refuses to perform, and in all artificial plantations it will always be a favourite addition to the woodland scene.

It is a fellow-countryman of the Latin Classic Poets, from whom it has received frequent and honourable mention. Even with us, it attains a considerable size; but in the milder climates of Italy, Spain, &c., it becomes a large tree, and reaches an age equal to that of some of our most venerable Oaks. Hence it not unfrequently acquired an historical interest; and for this reason perhaps, more than for its picturesque beauty, it was made the theme of poetic song. The Roman naturalist, Pliny, who flourished in the first century of the Christian Era, mentions a tree growing in the Vatican, which claimed a higher antiquity than Rome itself. It had brazen letters in the ancient Etruscan characters affixed to its trunk, from which it would appear, that before the Roman name was known it was a sacred tree. Its age must therefore have been 800 years at least. Three others are mentioned by the same author, growing at Tibur, which tradition made to be older than Tiburtus, who founded that city 1200 years B.C. Lowth considers the *Teil-tree* of Scripture to be identical with the Ilex, which abounds in many parts of Palestine; and it is more than probable that the Oak of the Holy Scriptures is either this, or some allied species of Quercus.

The Ilex was introduced into England previously to 1580; but it was then a great rarity, and little thought of. In Italy it is the prevailing evergreen, and in Sicily it abounds on the hills and all along the coast, ascending Mount Etna to an elevation of 3200 feet. It is easily propagated from the acorn, but is very impatient of being transplanted, owing to its sending its long roots perpendicularly downwards, which are furnished with but few rootlets, save at the extremities, and if these are injured, the young plant dies. This difficulty is obviated by sowing the acorns either in the spot where the trees are destined to stand, or by confining the roots in pots until they are required for planting. During their early stage they grow with considerable rapidity, but afterwards increase much more slowly. The bark is even, and of a light colour; the leaves of a dark bluish-green above, and more or less downy beneath, the younger shoots being as remarkable for their light hue as the full-grown tree is for the characteristic sombreness of its foliage. The shape of the leaf varies greatly in different individuals, and even not unfrequently on the same tree, being sometimes scarcely notched at all, at other times deeply serrated, and at others quite prickly. It is this last variety which has procured for it the name of "Holm Oak." It also resembles the Holm or Holly-tree, in having its most prickly leaves on the lowest branches. The acorn, which does not arrive at perfection until the second year, resembles that of the Oak, but is somewhat more slender, and the cup is scaly. Some trees bear sweet and edible acorns; those produced by others are bitter, and both kinds are sometimes to be found on the same tree. An allied species, *Quercus gramuntia*, which is so like the Ilex as to have been thought formerly merely a variety of the same tree, bears acorns, which when in perfection are as good as a chestnut, or even superior to it. These, according to Capt. S. C. Cook, are "the edible acorns of the ancients, which they believed fattened the tunny fish on their passage from the ocean to the Mediterranean; a fable, only proving that the acorns grew on the delicious shores and rocks of Andalusia, which, unhappily, is no longer the case. I have frequently seen them produced by individuals, and offered to the company, as *bon-bons* are in some countries, with a sort of apology for their small intrinsic value."

The wood of the Ilex is dark, close-grained, heavy, and very hard. It is also very durable and flexible, and, says Evelyn, "is serviceable for many uses, as stocks of tools, mallet-heads, mall-balls, chairs, axle-trees, wedges, beetles, pins, and, above all, for palisadoes, and in fortifications. Besides, it affords so good fuel, that it supplies all Spain almost with the best and most lasting of charcoals in vast abundance." Modern writers on the subject confirm this account, and recommend also its employment in ship-building.

The largest Ilex in the vicinity of London stands in the garden of Fulham Palace, and was planted probably by Bishop Compton; but this tree attains its greatest size in the south of England. Loudon mentions a tree at Mamhead in Devonshire, eighty-five feet high, the circumference of the trunk being eleven feet; and another at the same place fifty-five feet high, with a trunk twenty-two feet in circumference. At Mount Edgecumbe, in Devonshire, there are some very beautiful Ilex-trees growing within a few hundred yards of the sea-shore; and at Clowance, in Cornwall, there is a splendid group of these trees, the largest of which measures nine and a half feet in circumference at three feet from the ground: it there divides into two branches, one of which is six feet in circumference, the other five. Planted in groups as they here are, they are beautiful not only from the striking contrast which, both in summer and winter, they afford to the surrounding trees, but form a broad and important feature in the landscape.

At Trelowarren also in Cornwall are a number of finely-grown trees, varying from nine to twelve feet in circumference. The Ilexes at St. Michael's Mount, which Loudon describes as making a very fine appearance, must have greatly declined of late. They would now scarcely attract the passing notice of a visitor.

P. 118.

SYCAMORE AT KIPPENCROSS.

THE SYCAMORE.

ACER PSEUDO-PLATANUS.

Natural order—ACERINEÆ.

Class—OCTANDRIA. *Order*--MONOGYNIA.

IF in my history of forest trees I were to confine myself to those which are universally acknowledged to be indigenous to Britain, I should soon bring my labours to a close. England, though once a well-wooded country, never probably could boast of containing within it any great variety of species. The Oak, fortunately, no one thinks of denying to be our fellow-countryman: if any one were bold enough to do so, we could easily refute him by pointing to living trees older than any of our national records; or, if that did not suffice, to trunks of trees preserved in peat bogs, which were prostrated on their native soil centuries, probably, before the acorns were planted from which any trees now living sprung. But this is not the case with the Sycamore. No writer on the subject, so far as I can learn, looks on this tree in any other light than as a foreigner, but as a foreigner naturalized so completely that it will continue to sow its own seeds and nurse its own offspring, as long as England exists. The Oak, indeed, has greater right to claim an indigenous origin than we ourselves. There can be little doubt that the Oaks which now stock our forests, or convey our sailors to every region of the world, are lineal descendants of the first trees which ever grew in our island.

The Oak, on account of the age and size which it attains, the share which it had in the religious worship of our forefathers, its picturesque beauty, and its intimate connexion with naval architecture, is confessedly the most interesting of all the trees which grow in Britain. But the Sycamore is sadly deficient in these respects. It has neither extraordinary magnitude or longevity to recommend it. It was not contemporary in this country with the worshippers of trees; and I know not that it ever laid claim to be mentioned in connection with any national boast. It has even been denied the possession of any picturesque beauty. Evelyn says of it, "The Sycamore is much more in reputation for its shade than it deserves; for the honey-dew leaves, which fall early, like those of the Ash, turn to mucilage and noxious insects, and putrefy with the first moisture of the season; and are therefore, by my consent, to be banished from all curious gardens and avenues." If the trees, however, "be very tall and handsome, they are the more tolerable for distant walks, especially where other better trees prosper not so well, or where a sudden shade is expected. Some commend them to thicken copses, especially in parks, as least apt to the spoil of deer, and that it is good firewood."

With me, however, I confess, it is somewhat of a favourite. Its buds are of a very elegant shape, particularly when beginning to burst. When it grows in sheltered hedges the young shoots which spring from the base of the trunk, expand their pink or crimson leaves very early in the season, long before the tree has assumed its general foliage. In this stage they contrast beautifully with the dark blue of the violet and the delicate yellow of the primrose, and as a few sprigs may always be found when these favourite flowers are in their prime, they have invariably had a place in my early spring nosegays. Another reason why I regard it with peculiar good will is, that several trees of this species which I assisted to plant, at a time when they were no larger than a walking-cane, have now become stout, shady trees, though not yet twenty years old. But besides this, with its large and abundant leaves it forms a delightful shady retreat during the summer months: in the spring its graceful pendant clusters of flowers, diligently explored by bees and countless other insects, are among the most interesting natural objects of the season: while in autumn its tassels of winged seeds cannot fail to suggest pleasing and instructive reflections on the wise superintending Providence of the Almighty.

Gilpin speaks of the Sycamore in a less condemnatory tone than Evelyn. "The Great Maple," he says, "commonly called the Sycamore, is a grander and nobler tree than the smaller Maple; but it wants its elegance: it is coarse in proportion to its bulk. It forms, however, an impenetrable shade, and often receives well contrasted masses of light. Its bark has not the furrowed roughness of the Oak; but it has a species of roughness very picturesque. In itself it is smooth; but it peels off in large flakes, like the Planes, (to which, in other respects, it bears a near resemblance,) leaving patches of different hues, seams, and cracks, which are often picturesque."

According to Lauder, the Sycamore is a great favourite in Scotland, and is much planted about old aristocratic residences in that country. "The spring tints of the Sycamore," he says, "are rich, tender, glowing, and harmonious. In summer its deep green hue well accords with its grand and massive form; and the browns and dingy reds of its autumnal tints harmonize well with the other colours of the mixed grove, to which they give a fine depth of tone."

Having thus endeavoured to enlist the prejudices of my readers in favour of the Sycamore, I will proceed to describe it.

The name *Acer*, given to it by the Romans, is derived from *Acer, acris*, sharp or hard, (*ac*, Celtic, a point,) on account of the hardness of the wood, which was used for making spears and other sharp-pointed instruments; or, as some are pleased to say, from *acre ingenium*, a "*sharp wit*," from its being so much in use by the most ingenious artificers in fine works. Its specific name, *Pseudo-Platanus*, means *Mock-Plane*, being given to it in consequence of the resemblance borne by its leaves to those of the Plane-tree. The name Sycamore was given to it by the older Botanists, who erroneously believed it to be identical with the Sycamore,* or Mulberry-fig, of Palestine, which it somewhat resembles in the size and form of its leaves.

No tree propagates itself more readily in this

* From *syke*, a fig, and *moros*, a mulberry; being said to resemble the mulberry-tree in the leaf, and the fig in its fruit.

country, as may easily be inferred from the great number of seedlings which are to be found springing spontaneously from the ground in the vicinity of Sycamores which have begun to bear seeds. In its earliest stage, it is a puny herbaceous plant, furnished with two, or sometimes more, narrow smooth leaves entire at the edges; these are the cotyledonous leaves. Shortly afterwards, (for during the whole of its existence it is a rapid grower,) a few pointed and notched leaves, tinged with pink, are produced in the centre of these; and as the nursling increases in size, others appear, having the five-pointed unequally-notched lobes which characterize the matured foliage of the tree. At the end of a year it will have attained, under favourable circumstances, the height of eighteen inches. As a sapling it is remarkable for its straight growth, smooth purplish-brown bark, and large leaf-buds. In this stage of its growth it is a great favourite with school-boys, who, in the spring, when the sap begins to rise, slip off a cylinder of bark, and by removing a portion of the pith and wood, manufacture the shrill and unmusical instrument, a whistle. It produces flowers before it is twenty years old, but does not generally perfect its seeds until it has attained at least that age. In fifty or sixty years it reaches its full growth, and in the course of thirty or forty years more, thoroughly ripens its wood.

The leaves of the Sycamore in autumn are frequently observed to be covered with dark-coloured spots. This appearance is produced by numerous blackish fungi (*xyloma acerinum*), which, as soon as the first sharp frost has scattered the leaves on the ground, commence their office of converting the now useless vegetable substance into rich mould. At all periods of its growth its leaves are liable to be covered with a peculiar viscid substance, termed *honey-dew*, the origin of which has by some been attributed to insects, by others to the plant itself. It is now, I believe, generally admitted that the formation of this clammy sweet juice is to be assigned neither to the effect of disease in the plant, nor to the agency of insects: but, like the manna of the Ash and the gum of the Cistus, is to be considered as a natural exudation of the juices of the plant; what purpose it serves is, however, unknown. It is the presence of this substance which causes the unsightly appearance complained of by Evelyn, greatly aggravated as it is in the neighbourhood of inhabited houses by particles of soot floating in the air which rest on the leaves, and are detained there; as well as by the exuviæ of insects flocking to them for a repast.* A writer, quoted by Loudon, says, that the bees are so fond of the juice which exudes from the leaves of *Acer Platanoides*, that it would be worth while to plant the tree in the neighbourhood of places where hives are kept.

Thus is it that Nature is her own handmaid. The superfluous juices of the Sycamore are not lost, but are deposited on the surface of the leaf to afford a plentiful banquet to the tiny myriads that "wind their sultry horn" around us. Nothing is created in vain. We applaud the sentiment of the poet who sings of the flower that

* When this honey-dew is very abundant, it is liable to drop on any shrubs beneath (such as box, holly, &c.), and to turn their leaves black. The branches of such shrubs have been observed to be much infested with lichens.

"wastes its fragrance on the desert air;" but this is only because we do not know what end is answered either by its brilliant colouring or delicious perfume. Possibly, if we had a fuller and firmer faith in the unlimited Providence of God, we should believe that the flower secluded in the depths of some untrodden forest, is furnished with a symmetry and tinting and fragrance beyond all that human art can attain, for some purpose which could not so well be answered by any other means. As it is, we wonder that so much perfection should exist, seemingly all in vain; but it were, methinks, a more pious employment, if not so poetical, to admire and be thankful for all that we have been permitted to comprehend; and where limits are set to our apprehension, humbly and meekly to adore. But in truth, we have need of as much faith in natural as in revealed religion. We must "consider the lilies of the field" not only so far as they are within the cognizance of our senses; we must be content to lose ourselves in the devotional thoughts which force themselves upon us when the objects of sense are exhausted:

> "Thoughts which do often lie too deep for tears."

In May, before the leaves are thoroughly expanded, the Sycamore puts forth its elegant drooping clusters of green flowers, when the bee may be observed climbing about, and closely peering into, every opening bud. This insect is much indebted to the Sycamore, since its flowers, which abound in honey, not only are very numerous, but appear at a season when the supply of honey-bearing flowers is limited.

As soon as the flower is withered and has fallen off, the seed-vessels enlarge and acquire a reddish hue, which indeed in the autumn characterizes the whole tree.

"Nor unnoticed pass
The Sycamore, capricious in attire;
Now green, now tawny, and ere autumn yet
Has changed the woods, in scarlet honours bright."
COWPER.

Each of the two or three seed-vessels which succeed every flower is furnished with a membranous diverging wing, and it is owing to the presence of this that so many young plants may be discovered in the spring at a considerable distance from the

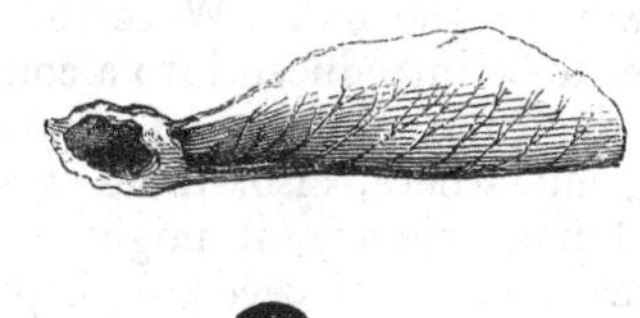

parent tree. When they have acquired their full size, which is early in autumn, they form clusters sufficiently large and conspicuous to alter the pictorial character of the tree. They do not fall off when ripe like acorns, chestnuts and other heavy seeds, but remain attached to the branches till the equinoctial gales set in, which serve the double purpose of separating them from the stalks and carrying them to some convenient place of growth. If, however, from growing in a sheltered spot, or from any other cause, they still retain their position, an event which frequently occurs, the seed-stalk rots from the effects of the winter's rain; and the violent winds which accompany the succeeding vernal equinox do not fail to deposit the majority of the seeds in a place well adapted for their growth, in full time to receive all the advantages of the genial season which follows. The seed itself is well protected against the severest vicissitudes of weather, first by the horny, or almost woody, case in which it is enclosed; and secondly, by the copious, soft, and glossy down which lines the seed-vessels, a covering alike impervious to cold and wet.

It may be, that many trees which have been introduced into a strange country, fail to propagate themselves extensively, because the attendant circumstances are not the same in the new country that they were in the old. Were the Sycamore, for instance, to be introduced into a country where no such periodical recurrence of rain and storms took place, and where, also, there was no interference of human agency, it might soon become extinct, inasmuch as its seeds, if kept dry for a year, generally lose their vegetative power. The Oak, if planted in a country uninhabited by man, and where no such friendly depredator as the rook or the squirrel acted the part of a skilful forester, would soon disappear. Its acorns would indeed fall to the ground, and perhaps germinate, but would rarely become trees, for the Oak, like many other trees, will not flourish under the shade of its own species. I may here observe, that the mast-bearing trees generally, such as the Oak, the Chestnut, and the Beech, are indebted for their propagation to animals whose instinct leads them to bury their food: those provided with winged seeds, such as the Sycamore, the Ash, and the Elm, to storms and tempests; and the drupe-bearing trees, (those, namely, which are furnished with stone-fruit,) to frugivorous birds, which fly away with the fruit and drop the seed.

Thus by the wise arrangement of the Almighty

do these several classes of trees derive the greatest benefit from what we, at first sight, might imagine to be most productive of injury.

From the extreme fecundity of this tree, Martyn argues that if it were truly indigenous, it would ere this have filled the whole country, instead of being a simple occupant of plantations and hedges. In Switzerland, Germany, Austria, and Italy, it is found abundantly in the mountainous forests, and may therefore with propriety be considered a native of those countries, whence it was probably introduced into Britain, in the end of the fifteenth or beginning of the sixteenth century. There are several varieties of Sycamore, which are propagated by grafting. The most remarkable among these are, the Yellow-leaved, or Costorphine Plane,* which is not common, except in the neighbourhood of the place from which it takes its name; and the Purple-leaved, so called from having the under surface of its leaves, especially in spring, tinged with dark purple. The value of all these, as ornamental trees, is much enhanced by the earliness of the season when they come into leaf.

Chaucer speaks of it as a rare exotic, in the fourteenth century. Gerard, in 1597, says, "The Great Maple is a stranger in England, only it groweth in the walkes and places of pleasure of noblemen, where it especially is planted for the shadowe-sake, and under the name of Sycamore tree." Parkinson, speaking of the same in 1640, says: "It is no where found, wild or natural, in our land, that I can learn; but only planted in

* In Scotland, the Sycamore is frequently called "The Plane."

orchards or walkes for the shadowes sake." It abounds in sweet juice, of which, says Evelyn, "the tree being wounded, in a short time yields sufficient quantity to brew with, so as with one bushel of malt is made as good ale as with four bushels with ordinary water." According to Sir T. Dick Lauder, "The Sycamore has been proved to be capable of yielding sugar: Incisions were made, at five feet from the ground, in the bark of a tree of this species, about forty-five years old. A colourless and transparent sap flowed freely, so as in two or three hours to fill a bottle capable of containing a pound of water. Three bottles and a half were collected, weighing in all three pounds, four ounces. The sap was evaporated by the heat of a fire, and gave two hundred and fourteen grains of a product, in colour resembling raw sugar; and sweet in taste, with a peculiar flavour. After being kept fifteen months, this sugar was slightly moist on the surface. The quantity of sap employed in the evaporation was 24,960 grains, from which 214 grains of sugar were obtained; therefore, 116 parts of sap yielded one part of sugar."

An allied species, *Acer Saccharinum*, or Sugar Maple, which is found in great quantities in Canada, New Brunswick, Nova Scotia, and other parts of North America, yields a similar saccharine juice, in such quantities that maple-sugar is an important article of manufacture. It has been computed, that in the northern parts of the two states of New York and Pennsylvania, there are ten millions of acres which produce these trees in the proportion of thirty to an acre. The season for tapping is in February and March, while the cold continues intense, and the snow is still on the ground. A tree of ordinary size yields from fifteen to thirty gallons of sap, from which are made from two to four pounds of sugar. The tree is not at all injured by the operation, but continues to flourish, after having been annually tapped for forty years without intermission. The produce is consumed principally in the neighbourhood of the place where it is manufactured;* the sugar from the cane being preferred whenever it can be readily procured.

Our Sycamore is not sufficiently productive of sugar to be ever employed in this way, even if the manufacture were legalized; but it is by no means a worthless tree. Its wood was much used for making platters before earthenware plates were generally introduced, and in rural districts is still applied to the same purpose. When the tree is young the wood is white, but acquires a yellow or brown hue as it increases in age. It is close-grained, but not hard, and does not readily warp, and, being easily worked either by the hand or lathe, was formerly held in high estimation for the purpose above mentioned. It is sought by the joiner and cabinet maker, and is also used for making musical instruments and cider-screws. It forms also a very valuable fuel, burning slowly and giving out a great deal of heat. Not only on account of its uses in the arts and manufactures, and its dense foliage in summer, was its growth encouraged; but it was planted in the vicinity of houses, from the supposition that it was the Sycamore of Scripture; this however is not the case, the tree

* It is, however, stated that ten millions of pounds are annually imported into the United States.

into which Zacchæus climbed to see our Saviour pass on his way to Jerusalem being the *Ficus Sycomorus*. However, as the error once generally prevailed, both that tree and our tree bearing the same name have been selected by the inventors of the language of flowers to indicate curiosity.

Dr. Shaw, speaking of the Sycamore of the East, says, "The mummy-chests, and whatever figures and instruments of wood are found in the catacombs, are all of them of Sycamore, which, though spongy and porous to appearance, has, notwithstanding, continued entire and uncorrupted for at least three thousand years."

"From its value in furnishing wood for various uses, from the grateful shade which its wide-spreading branches afforded, and on account of the fruit, which, Mallet says, the Egyptians live upon and hold in the highest estimation, we perceive the loss which the ancient inhabitants must have felt 'when their vines were destroyed with hail, and their Sycamore-trees with frost.'"*

"The Great Maple, or European Sycamore, will grow in any soil not saturated with moisture; but it seems to prefer one that is dry and free, rather than one that is stiff and moist. It will grow in exposed situations, and especially on the sea coast, and maintain its erect position against the sea breeze, better than most other trees. It is in use for this purpose in Scotland, and also for planting round farm-houses and cottages on the bleak hills. In such situations, an instance can hardly be found of the head of the tree leaning more to one side than another.

* Ps. lxxviii. 47.

Even when the wind blows strongly in one direction for nine months in the year, this tree maintains its perpendicularity and symmetrical form." —*Loudon.*

Though a fast grower, the Sycamore does not attain a remarkably large size, and it is as little noted for its longevity. It does not materially increase in size after having reached the age of sixty years, but requires from thirty to forty years more to bring its timber to perfection.

At the age of from one hundred and fifty to two hundred years, it usually closes its term of life; though much older trees are on record.

Sir Thomas Dick Lauder mentions "a Sycamore at Calder House, in the county of Edinburgh, standing on the pleasure ground, on the road from the house to the church, which, on the 4th of October, 1799, measured seventeen feet, seven inches in girth; at the ground it measured twenty feet three inches. Its trunk is twelve feet high, and it then divides into five great arms. Its branches extend in diameter about sixty feet. This tree is known to have been planted before the Reformation, and is therefore not less than three hundred years old; yet it has the appearance of being perfectly sound. It was the tree to which, long ago, the iron jugs (a species of pillory) were fastened. The tree came gradually to grow over them, and they have now been completely enclosed in its trunk for a considerable time. At the place where they are enclosed, there is a great protuberance on the south side of the tree, at the height of between four and five feet.

"A Sycamore at Newbottle Abbey, situated north-west from the house, and the largest tree of its kind about the place, in 1789, measured, at four feet from the roots, eighteen feet seven inches in girth. At the height of two feet and a half from the ground, it was twenty-four feet four inches, and it is about seventy feet high. It has the appearance of great antiquity, but seems still to be sound. Many other Sycamores at Newbottle were planted before the Reformation, and apparently about the same time with this, though they are inferior in size. This tree was probably planted before the year 1530."

"The Sycamore at Kippencross is truly a noble tree. It has been figured by Nattes in his 'Scotia Depicta.' He states it to have been, in 1801, twenty-eight feet nine inches in girth, with a stem of thirty feet. He must have measured its circumference at the ground, as, when taken breast high, in 1798, its girth was only twenty-two feet six inches. In 1809, this tree was in full health and beauty."* All that is known of the age of this tree is, that in Charles the Second's reign it went by the name of "The big tree of Kippencross."

* *Lauder's Gilpin.*

Pages 124—126.

MAPLE IN BOLDRE CHURCH-YARD.

THE COMMON, OR FIELD MAPLE.

ACER CAMPESTRE.

Natural order.—ACERINEÆ.

Class—OCTANDRIA. *Order*—MONOGYNIA.

THOUGH the tree last described is much larger and more generally known than the present species, it has so long universally borne the name of Sycamore, that the generic name of "Maple" is now almost exclusively applied to the smaller tree, the only species, in fact, which is indigenous to this country. Many persons probably are not aware, that the two trees belong to the same family, for if we except the keys, or clusters of winged seeds, they have to the casual observer few points of resemblance.

The Sycamore justly claims the right of being considered a large tree: the circumference of its trunk is considerable; it frequently covers a wide space of ground with its spreading limbs; it casts a dense shade, and its leaves exceed in size those of most of our common trees. But the Maple rarely attains a size which entitles it to be considered a tree at all; its foliage is meagre and unpretending, while its value in hedge-making induces its owners to preserve as much as possible its character of an overgrown shrub. Such, accordingly, we generally find it when it grows in hedges; and when met with among other trees it is mostly as underwood. Its leaves, like those of the Sycamore, are five-lobed, but obtuse and much smaller. Its flowers, which appear in April,

about a fortnight before the leaves and abound in saccharine juice, are similarly constructed with those of the Sycamore, but grow in erect, instead of drooping, clusters; and the keys, which differ principally as it regards size from those of the other species, are tinged with red. Besides being indigenous to Britain, it grows naturally in the middle and south of the European Continent, and in the north of Asia.

In France, it appears to serve the purposes of man more than in this country. According to

Loudon, "The young shoots, being tough and flexible, are employed by the coachmen in some parts of France instead of whips. The tree is much used in the same country for forming hedges, and for filling up gaps in old fences. It is also employed in topiary works, in geometrical gardens, its branches being found to bear the shears better than those of most other trees. The leaves and young shoots are gathered green, and dried for winter provender for cattle. The sap yields more sugar, in proportion to the quantity taken, than that of the Sycamore; but the tree does not bleed freely. In Britain, the tree is seldom planted for any other purpose than that of ornament, in which it is effective, by adding to the variety of a collection, rather than to its positive beauty." The wood makes excellent fuel, and the very best charcoal. Evelyn says of it, "By shredding up the boughs to a head, I have caused it to shoot to a considerable height in a little time; but if you will lop it for the fire, let it be done in January; and indeed it is observed to be of noxious influence to subnascent plants of other kinds, by reason of a clammy dew, which it sheds upon them, and therefore they should not be indulged in pollards, or spreading trees, but to thicken underwoods and copses. The timber is far superior to Beech for all purposes of the turner, who seeks it for dishes, cups, trays, trenchers, &c., as the joiner for tables, inlayings, and for the delicateness of the grain, when the knurs and nodosities are rarely diapered, which does but advance its price: our turners will work it so thin, that it is almost transparent."

As an ornament to the landscape, the Maple

has not much to recommend it. Gilpin says of it, "The Maple is an uncommon tree, though a common bush. Its wood is of little value; and it is therefore rarely suffered to increase. We seldom see it employed in any nobler service than in filling up its part in a hedge, in company with thorns, and briars, and other ditch trumpery." And although he afterwards says, "In the few instances I have met with of this tree in a state of maturity, its form has appeared picturesque;" yet his praise of it is so exceedingly slight, that I have very little doubt that his eye, acute as it was to discern what is beautiful in the general features of nature, could have alighted with greater pleasure on almost any other kind of tree that can be named. Nevertheless, he has given to the Maple a deeper interest than it ever possessed before; for "under the large Maple in Boldre churchyard, the Rev. W. Gilpin, after fulfilling his duties in the most exemplary manner for twenty years, as rector of this parish of Boldre, chose for his last resting place this sweet sequestered spot, amidst the scenes he so much loved, and has so well described." (*Strutt.*)

By the ancients hardly any wood was more valued than that of the Maple, in so much, that Virgil represents one of his kings as seated on a Maple throne. The great naturalist Pliny, says that its trunk, for beauty and firmness of grain, is inferior only to the Citron-wood. One kind, from the varied character of its veining, was named the Peacock Maple. The knots called *Brusca* and *Mollusca*, were most valued and manufactured into various ornaments which the limited size of the material would allow.

In the Molluscum the veins were wide apart from each other. The Bruscum was deemed most valuable, when the arrangement of the veins resembled some animal, (as was occasionally the case,) and gave the wood a dark hue. The latter was preferred for making tables. "And such spotted tables," says Evelyn, "were the famous Tigrin and Pantherine curiosities; not so called from being supported with figures carved like those beasts, as some conceive, and was in use even in our grandfathers' days, but from the natural spots and maculations. Such a table was that of Cicero, which cost him ten thousand sesterces (about 62*l.* sterling); such another had Asinius Gallus. That of King Juba was sold for fifteen thousand; and another which I read of, valued at a hundred and forty thousand sesterces, which, at about three halfpence sterling, arrives to a pretty sum (875*l.* sterling); and yet that of Mauritanian Ptolemie was far richer, containing four feet and a half diameter, three inches thick, which is reported to have sold for its weight in gold. Of that value they were, and so madly luxurious the age, that when they at any time reproached their wives for their wanton expensiveness in pearl and other rich trifles, they were wont to retort, and turn the *tables* upon their husbands."*

Spenser appears to have considered the timber of the standing tree peculiarly liable to decay, for he speaks of

"The Maple seldom inward sound."

The largest Maple now existing in Britain, and the only one to which any particular interest at-

* The Bird's-eye Maple of modern cabinet-makers is the wood of the Sugar, or Rock, Maple. The trunk of this tree is rejected for

taches itself, is that mentioned above, as overshadowing the grave of Gilpin in Boldre churchyard. It is ten feet in circumference at the ground, and at four feet from the ground, is seven feet six inches. The trunk divides into branches at twelve feet, and the entire height of the tree is forty-five feet.

The following notice of the Maple is extracted from the *Journal of a Naturalist.* "The Maple is found growing in all our fences, generally reduced by the hedger's bill, to serve the same humble purposes as the thorns and sloes associated with it. Sometimes, however, it is permitted to assume the rank of a tree, when, if not possessing dignity, it is certainly beautiful, and becomes an ornament in the hedge-row. It is the earliest sylvan beau that is weary of its summer suit; first shifting its dress to ochrey shades, then trying a deeper tint, and lastly, assuming an orange vest; thus setting a fashion that ere long becomes the garb of all except the rustic oak, which looks regardlessly at the beau, and keeps its verdant robe unchanged.

"Soon tired of this, the Maple takes a pattern from his sober neighbour Ash, throws its gaudy trim away, and patiently awaits with all his peers the next new change. In spring, the woodbine wreaths its knots of green around the rugged limbs of the Maple; the rose beneath puts on its emerald gems, and then our gallant sir will wear such colours too, fluttering through all its summer's day. When first the Maple begins to au-

civil and naval architecture ; but the wood of old trees is valued for inlaying mahogany. The appearance from which it derives its name proceeds from the twisting of the silver grain, which produces numerous knots like the eye of birds.

tumnize the grove, the extremities of the boughs alone change their colour, but all the internal and more sheltered parts still retain their verdure, which gives to the tree the effect of a great depth of shade, and displays advantageously the light, lively colouring of the sprays. We find the Maple useful in our hedges, not from the opposition it affords, but by reason of its very quick growth from the stool after it has been cut, whence it makes a fence in a shorter time than most of its companions; and when firewood is an object, it soon becomes sufficiently large for this purpose. The singular ruggedness of the branches and shoots, when they have attained a year's growth, and the depth of the furrows, give it a strongly marked character among our shrubs. The under side of the leaves in autumn, when they become yellow, dashed here and there with a few specks of red and brown, appear, when magnified, like a very beautiful and perfect mosaic pavement, with all its tesseræ arranged and fitted. If one of these rugged young shoots be cut through horizontally with a sharp knife, its cork-like bark presents the figure of a star with five or more rays, sometimes irregularly, but generally exactly defined. A thin slice from this surface is a beautiful and curious object in the microscope; exhibiting the different channels, and variously-formed tubes, through which the sap flows, and the air circulates for the supply of all the diversified requirements of the plant. And it is good and delightful to contemplate the wonderful mechanism that has been devised by the Almighty Architect, for the sustenance and particular necessities of the simple Maple, this "ditch trumpery," as Gilpin calls it;

which naturally leads one to consider, that, if He had so regarded such humble objects, how much more has He counted worthy of His beneficence, the more highly destined orders of His creation!"

THE ASH.

Fraxinus excelsior.

Natural Order—Jasmineæ.

Class—Diandria. *Order*—Monogynia.

The Ash is, in utility, inferior only to the Oak, and like that tree an undoubted child of the soil. Not remarkable for robustness, grandeur, or longevity, it rests its claims on qualities scarcely less striking. In height, gracefulness of form, and elegance of foliage, it has no superiors, scarcely any competitor. Its favourite haunts, too, give it an additional charm.

Far away in some secluded valley, through which a mountain stream, prolific in miniature waterfalls, hurries or lingers,

"—at its own sweet will,"

now penned up between party-coloured rocks, and now undermining the deep alluvial soil, which, in furtherance of the end for which it bubbled forth from the earth, it brought with it, ages ago, from the hills—among straggling mosses and strange looking liverworts, which have no name save in the books and memory of the Naturalist—here the Ash is in its home. You may find it at times, a handsome looking tree in the neighbourhood of farms or in parks, and contributing greatly to the beauty of the landscape even in these localities; or you may see it lifting aloft a wretched broom-like head on a pale, disfigured stem, maimed and scarred throughout its whole length by the axe of the hard-handed farmer, who dreads the noxious influence of its drippings on his meadows and corn-lands; but neither of these is the tree which the lover of nature pictures to himself when he questions himself on his recollection of the Ash. This must be a tree that *enjoys*, in common with many of its brethren, the beauties of the haunts I have described, not simply living and flourishing, but actually delighting in the brilliant sparkling of the water, watching the ousel as he bathes in his rapid flight, gracefully sweeping its branches over the stream, climbing up the sides of the steep hill, or endeavouring to peep at what is passing in the world beyond. This is "the Ash" of the *Poet* and the *Painter*, and something of these the true *Naturalist* must be, though he is neither cunning to touch the lyre or handle the pencil—though unable in any way to give expression to his thoughts.

> ———"Here amid the brook,
> Grey as the stone to which it clung, half root,
> Half trunk, the young Ash rises from the rock:
> And there the parent lifts its lofty head,
> And spreads its graceful boughs; the passing wind
> With twinkling motion lifts the silent leaves,
> And shakes its rattling tufts."
>
> *Southey.*

The Ash was well known to the Greeks, who called it *melea*. Homer arms his heroes with an ashen spear, and Cupid's arrows were originally made of the same wood, though he afterwards stood indebted to a less cheerful tree, the Cypress.

The Romans called it *Fraxinus*, a name which naturalists still retain, but the derivation of which is very uncertain. They employed its wood in the manufacture of weapons and many kinds of agricultural implements. In the Teutonic Mythology, the Ash holds a conspicuous place. Under the shade of an enormous tree, of which the branches overspread the earth, the top reached to Heaven, and the roots to the infernal regions, the gods held their court. On the summit was perched an eagle, who watched the course of all earthly affairs, assisted by a squirrel, who employed his time in descending and ascending to examine into, and report upon, what was passing beneath. Pliny gravely informs us that the serpent would rather creep into the fire than shelter itself in its branches:* and Dioscorides, the physician, states that the juice of the Ash is an antidote against the bite of the same reptile.

But we need not go back to ages so remote as these for superstitious opinions respecting this tree. Gilbert White, in his classical history of Selborne, says: "In a farm-yard, near the middle of this village, stands at this day, a row of pollard-ashes, which, by the seams and long cicatrices down their sides, manifestly show that in former times they have been cleft asunder. These trees, when young and flexible, were severed and held open by wedges, while diseased children, stripped naked, were pushed through the apertures, under a persuasion, that by such a process the poor babes would be cured of their infirmity.

* "This," says Evelyn, "is an old imposture of Pliny, who either took it upon trust, or else we mistake the tree."

As soon as the operation was over, the tree in the suffering part was plastered with loam, and carefully swathed up. If the parts coalesced and soldered together, as usually fell out, where the feat was performed with any adroitness at all, the party was cured; but where the cleft continued to gape, the operation, it was supposed, would prove ineffectual. Having occasion to enlarge my garden not long since, I cut down two or three such trees, one of which did not grow together. We have several persons now living in the village, who in their childhood were supposed to be healed by this superstitious ceremony, derived down, perhaps, from our Saxon ancestors, who practised it before their conversion to Christianity." The same custom was known to Evelyn, who half believes in the efficacy of the ceremony.* If we may credit Phillips, the present enlightened age is not exempt from the same silly belief. He says: "In the south-east part of the kingdom, the country people split young Ash trees, and make their distempered children pass through the chasm in hopes of a cure.† They have also a superstitious custom of boring a hole in an Ash, and fastening in a shrew mouse; a few strokes with a branch of this tree, is then accounted a sovereign remedy against cramp and lameness in cattle, which are ignorantly supposed to proceed from this harmless animal."‡ Such a tree was named from the unfortunate victim "a shrew-

* Hunter's Evelyn's Sylva, vol. i. p. 151.

† A writer in the Gardeners' Chronicle for April 1846, states, that there is now living in Sussex a man, who when an infant, about 50 years ago, was passed through an Ash tree, at Todhurst, as a remedy for hernia.

‡ Sylva Florifera, vol. i. p. 83.

ash." White thus describes one which about the middle of the last century stood in the village of Selborne: "At the south corner of the Plestor, or area, near the Church, there stood, about twenty years ago, a very old, grotesque, hollow, pollard-ash, which for ages had been looked on with no small veneration as a *shrew-ash.* Now, a shrew-ash is an Ash whose twigs or branches, when gently applied to the limbs of cattle, will immediately relieve the pains which a beast suffers from the running of a shrew mouse over the part affected: for it is supposed that a shrew-mouse is of so baleful and deleterious a nature, that wherever it creeps over a beast, be it horse, cow, or sheep, the suffering animal is afflicted with cruel anguish, and threatened with the loss of the use of the limb. Against this accident, to which they were continually liable, our provident forefathers always kept a shrew-ash at hand, which, when once medicated, would maintain its virtues for ever. A shrew-ash was made thus:—Into the body of the tree, a deep hole was bored with an auger, and a poor devoted shrew-mouse was thrust in alive, and plugged in, no doubt with several incantations, long since forgotten. As the ceremonies necessary for such a consecration are no longer understood, all succession is at an end, and no such tree is known to exist in the manor or hundred."

Lightfoot says, that in many parts of the Highlands of Scotland, at the birth of a child, the nurse puts one end of a great stick of this tree into the fire, and while it is burning, receives into a spoon the sap or juice which oozes out at the other end, and administers this as the first spoonful of food to the new-born infant.

The English name of this tree is derived from the Saxon *Æsc*. The common opinion that it is so called from the colour of its bark closely resembling that of wood-ashes is incorrect.

Gilpin's description of its general character is as usual accurate and truthful: "I have sometimes heard the Oak called the Hercules of the forest, and the Ash the Venus. The comparison is not amiss; for the Oak joins the ideas of strength and beauty; while the Ash rather joins the ideas of beauty and elegance. Virgil marks the character of the Ash as peculiarly beautiful:

"Fairest of forest-trees the Ash."*

The Ash generally carries its principal stem higher than the Oak, and rises in an easy, flowing, line. But its chief beauty consists in the lightness of its whole appearance. Its branches at first keep close to the trunk, and form acute angles with it: but as they begin to lengthen, they generally take an easy sweep; and the looseness of the leaves corresponding with the lightness of the spray, the whole forms an elegant depending foliage. Nothing can have a better effect than an old Ash hanging from the corner of a wood, and bringing off the heaviness of the other foliage with its loose, pendent branches. And yet in some soils, I have seen the Ash lose much of its beauty in the decline of age. Its foliage becomes rare and meagre; and its branches, instead of hanging loosely, often start away in disagreeable forms. In short, the Ash often loses that grandeur and beauty in old age, which the generality

* "Fraxinus in sylvis pulcherrima."

of trees, and particularly the Oak, preserve till a late period of their existence."

The Ash is indigenous throughout the greater part of Europe,* the north of Africa, and some parts of Asia. It rises freely from seed, and in favourable situations it grows rapidly. Its roots are remarkable for their tendency to take a horizontal direction, and, being abundantly furnished with fibres which approach closely to the surface, effectually check the growth of almost all other vegetation. Hence has originated the erroneous notion that the drippings of its leaves are peculiarly noxious. They dislike the presence of stagnant water; but delight to approach as closely as possible to the gravelly bed of a running stream. Owing to these instincts, if they may be so called, the Ash outstrips any other tree when it grows on the shallow rich soil which lines the course of our mountain streams. "It is by no means convenient to plant Ash in plow-lands, for the roots will be obnoxious to the coulter; and the shade of the tree is malignant both to corn and grass, when the head and branches over-drip and emaciate them.— The Ash delights in the best land, which it will soon impoverish, yet grows in any, so it be not over stiff, wet, and approaching to the marshy, unless it be first well drained: by the banks of

* Professor Jameson is disposed to think that in Scotland the Ash is not indigenous "The Ash and the Beech have a place in the *Flora Scotica* of Lightfoot and Hooker, and they have long ornamented our woods and plantations. But there is great reason to doubt their being truly indigenous to this country, or having formed any part of the ancient forests. No traces of them occur in our peat mosses: yet Ash-seeds and Beech-mast would in all probability have proved as indestructible as Hazel nuts or Fir cones, which are abundant in many peat mosses." (*Note in Jamesons's Journal.*)

sweet and crystal rivers and streams I have observed them to thrive infinitely."*

The young plants are readily distinguished from other saplings, in winter and early spring, by their ash-coloured tint, their remarkable, black buds, and the flattened or *compressed* shape of the twigs, a peculiarity which is most perceptible near the terminal pair of buds. In Summer, the leaves are a no less certain distinguishing character. They are technically termed *pinnate*, and are composed of about five pairs of acute, notched leaflets, with a terminal odd one, which last is occasionally abortive.

The foliage of the Ash is very late in making its appearance; consequently in early Spring it cannot compete in beauty with other forest trees which are less sluggish in donning their green attire. It is equally remarkable, too, for the earliness of the season at which it sheds its foliage.

"Its leaf is much tenderer than that of the oak, and sooner receives impressions from the winds and frost. Instead of contributing its tint, therefore, in the wane of the year, among the many coloured offspring of the woods, it shrinks from the blast, drops its leaf, and in each scene where it predominates, leaves wide blanks of desolated boughs, amidst foliage yet fresh and verdant. Before its decay, we sometimes see its leaf tinged with a fine yellow, well contrasted with the neighbouring greens. But this is one of Nature's casual beauties. Much oftener its leaf decays in a dark, muddy, unpleasing tint. And yet sometimes notwithstanding this early loss of its foli-

* Evelyn's Sylva.

age, we see the Ash, in a sheltered situation, when the rains have been abundant and the season mild, retain its green (a light pleasing green) when the oak and the elm in its neighbourhood have put on their autumnal attire."*

The precise time at which it sheds its leaves varies much in different individuals, and this difference arises not only from situation, but from other causes, for sometimes in the same hedgerow many trees will have cast their foliage while others show no symptom of decay.

"The shade of the Ash," says Evelyn, "is not to be endured, because the leaves produce a noxious insect; and for displaying themselves so very late, and falling very early, not to be planted for umbrage or ornament, especially near the garden, since (besides their predatious roots) the leaves dropping with so long a stalk, are drawn by clusters into the worm-holes, which foul the alleys with their keys, and suddenly infect the ground." For these reasons, landscape-gardeners do not recommend the extensive plantation of the Ash in localities where a permanent mass of foliage is desired, nor in the vicinity of mansions; nevertheless their growth may safely be encouraged "by the banks of sweet and crystal rivers and streams," according to Evelyn, their favourite haunts.

As to the uses of the foliage, Phillips says: "The Romans used the Ash-leaves for fodder, which were esteemed better for cattle than those of any other trees, the Elm excepted; and they were also used for the same purpose in this country, before agriculture was so well understood, and

* Gilpin.

our fields clothed with artificial grasses. In Queen Elizabeth's time, the inhabitants of Colten and Hawkshead fells remonstrated against the number of forges in the country, because they consumed all the loppings and croppings, which were the sole winter food for their cattle. In the north of Lancashire they still lop the Ash to feed the cattle in autumn when the grass is upon the decline, the cattle peeling off the bark." "Its leaf and rind are nutritive to deer, and much used in browzing them in summer. The keepers of the forest therefore seek out all the Ash-trees they can find, which are for this purpose mangled and destroyed."* Ash-leaves, as well as the leaves of the Sloe, are said to be used extensively for the purpose of adulterating tea, which they resemble in shape and size. It is somewhat consolatory to know that either substitute, if taken in moderation, is innocent.

In Britain, the Ash, in its living state, is not liable to be much disfigured by the attacks of insects; but on the Continent the case is very different, as will be seen from the following account quoted from various authors by Loudon in his valuable *Arboretum Britannicum:* "If nature had produced the Ash for no other purpose than for the embellishment of forests," says the writer of the article *Fraxinus* in the *Nouveau Du Hamel,* "we might almost say that she had failed in her end, or had opposed herself to her own views, in destining the leaves of that tree to be the food of an insect *Cantharis vesicatoria* (Spanish-fly or blister-fly), a beetle of a beautiful golden green, with black antennæ, which

* Gilpin.

devours them with avidity. The Ash is no sooner covered with leaves, than these are attacked by such a number of cantharides, or Spanish-flies, that the trees, during the remainder of the summer, have a dismal appearance; and, though the insect which devours the leaves may please the eye by its elegant form, and its colours of green and gold, yet it spreads abroad a smell which is so disagreeable, that it causes the common Ash to be excluded from our forests, where the flowering Ash, and some of the American species, are alone introduced! M. Pirolle, in one of the early volumes of the *Bon Jardinier*, mentions that, even when the cantharides are dead on the trees, they become dried to a powder, which it is difficult to pass the trees without inhaling. The particles of this powder, being parts of these flies that cause the blistering of the skin when a blister plaster is applied, are of course, dangerous to persons who inhale them; and on this account, Ash trees are never planted near villages in France." These insects being never numerous in England, there is no fear of any such results.

Gilpin's remarks on the spray of the Ash are well worth the attention of the artist. After pointing out the peculiar character of the Oak, he proceeds to say: "The spray of the Ash is very different. As the boughs of the Ash are less complex, so is its spray. Instead of the thick intermingled bushiness which the spray of the Oak exhibits, that of the Ash is much more simple, running in a kind of irregular parallels. The main stem holds its course, forming at the same time a beautiful sweep; but the spray does not divide, like that of the Oak, from the extremity of the last year's shoot, but springs from the side of it. Two shoots spring out opposite to each other, and each pair in a contrary direction. Rarely, however, both the shoots of either side come to maturity; one of them is commonly lost as the tree increases, or at least makes no appearance in comparison with the other, which takes the lead. So that notwithstanding this natural regularity of growth, so injurious to the beauty of the Spruce-fir, and some other trees, the Ash never contracts the least disgusting formality from it. It may even secure great picturesque beauty, for sometimes the whole branch is lost as far as one of the lateral shoots, and this occasions a kind of rectangular junction, which forms a beautiful contrast with the other spray, and gives an elegant mode of hanging to the tree.

"This points out another difference between the spray of the Oak and that of the Ash. The spray of the Oak seldom shoots out from the under sides of the larger branches; and it is this, together with the strength and firmness of the branches, which keeps them in a horizontal form. But the spray of the Ash as often breaks out on the under side as in the upper; and being of a texture weaker than that of the Oak, it generally, as the bough increases, depends below the larger branch, and rising again, forms, in full grown trees especially, very elegant pendent boughs."

This description is so very accurate and truthful, that the reader, if he is at all conversant with woodland scenery, can scarcely fail to recognise the portrait.

When the Ash has attained a considerable size, the spray assumes in early Spring, an appearance very different from that which characterised the younger tree. This is occasioned by the numerous clusters of flowers which appear at the extremities of the branches, at least a month before the leaves. These flowers are minute and remark-

ably simple in their structure, being destitute both of calyx and corolla: but being exceedingly numerous, and of a dark purple colour, they are very conspicuous, and add materially to the ordinary graceful character of the tree. They grow in dense clusters on the extremities of those branches which were produced in the former year: and, buried among them, lie the rudiments of the future leading shoot. They are difficult to describe except in the technical language of the botanist, but will amply reward any one who will take the pains to examine them closely: for, minute as they are, they are very elegant, and the rich purple contrasts beautifully with the delicate greenish-yellow tint of the flower stalks, though when the tree is observed from a distance, the latter are so closely concealed by the flowers as to be scarcely apparent. In its earlier stage of growth, the mass of unexpanded flowers is not unlike an irregularly granulated fruit; it eventually becomes diffuse, and is finally succeeded by bunches of pendent seeds, not inappropriately called *keys.** They differ from the keys of the Sycamore in growing singly, instead of in pairs, but like them are winged, and remain firmly attached to the tree, until the season when winds prevail sufficiently powerful to strip them from the branches, and carry them a considerable distance from the parent tree. How wise a provision this is, is very conspicuous in the case of the Ash, for, as we have seen above, the roots of this tree naturally extend horizontally so near the surface as to exhaust the soil and, consequently, to render it unfit for the nourishment of seedlings

* The Latins termed the seed of the Ash *lingua avis* (*bird's tongue*) from some fancied resemblance in shape.

of the same species. So firmly indeed are the keys attached to the twig, that not only may the tree be discriminated in winter by its bunches of brown seeds; but it is far from unusual to see the ragged remnants of the previous year mixed up with the fresh flowers and foliage.

The Ash is liable to a disease in its inflorescence which deserves to be noticed. It occurs in either particular trees or seasons; the whole flower-bud, without expanding at all, becomes an irregular, solid mass, and of course does not ripen its seeds.

It has been observed already, that the season at which the Ash sheds its leaves varies considerably in different individuals. It is also worthy of remark, that individual trees also vary greatly in the quantity of seeds produced, and that those which bear but few seeds compensate for their sterility, by a greater profusion of foliage, which they also retain until a much later period in the year. This phenomenon may be explained on the ground that when there is an abundant produce of seed, the tree reserves its energies in order to mature them, consequently the foliage is thrown off early in the Autumn: but when there is no such demand for the nourishment of seed, the tree expends all its vigour on the leaves, which are consequently numerous, and so healthy as to be little affected by the early frosts of Autumn.*

By the facility of transit which its winged

* My readers, if they have had any experience in gardening, must be well aware that this law applies to most, if not all, plants that come under their care. A *healthy* state of foliage is indispensable to the production of perfect flowers and fruit; anything more than this has a contrary effect; a superabundance of leaves, being usually attended by a defective produce of both flowers and fruit.

appendage affords to the seed of the Ash, we are to account for the appearance of trees in the very strange situations in which they are sometimes found, springing for instance, from church towers, ruins, and crags inaccessible to man. Dr. Plott, in his Natural History of Oxfordshire, mentions a singular instance of this vegetable waywardness; "An Ash-key rooted itself on a decayed willow, and finding, as it increased, a deficiency of nourishment in the mother plant, began to insinuate its fibres by degrees, through the trunk of the willow, into the earth. There, receiving an additional recruit, it began to thrive and expand itself to such a size, that it burst the willow in pieces, which fell away from it; and what was before the root of the Ash, being now exposed to the air, became the solid trunk of a vigorous tree."

Ash-keys were held in high repute by the ancient physicians for their medicinal properties. They were also preserved with salt and vinegar, and sent to table as a sauce, when, says Evelyn, "being pickled tender they afford a delicate salading."

From a foreign species of Ash, *Fraxinus Ornus*, of Linnæus, *Ornus Europœa* of modern authors, is procured a substance which, from its appearance somewhat according with the description of the miraculous food of the Israelites in the wilderness, is called Manna. "This substance is chiefly collected in Calabria and Sicily; where, according to the *Materia Medica* of Geoffroy, the manna runs of itself from the trunks of some trees, while it does not flow from others, unless wounds are made in the bark. Those trees which yield the manna spontaneously grow in the most favourable situations, and the sap runs from them spontaneously only during the greatest heats of summer.

It begins to ooze out about mid-day, in the form of a clear liquid, which soon thickens, and continues to appear until the cool of the evening; when it begins to harden into granules, which are scraped off the following morning. When the night has been damp or rainy, the manna does not harden, but runs to the ground and is lost. This kind is called manna in tears, or *manna lagrimi,* and is as white and pure as the finest sugar. About the end of July, when the liquid ceases to flow of itself, incisions are made through the bark and soft wood; and into these incisions slender pieces of straw, or twig, are inserted, on which the manna runs, and coating them over, hardens on them. This is the common manna of the shops, which is thus collected in the form of tubes, and is called *manna in cannoli.* Another, and inferior sort, is procured by making an oblong incision in the trees in July or August, and taking off a piece of the bark about three inches in length, and two inches in breadth. This kind, which is called *manna-grass,* is the coarsest; but as it is produced with less trouble, it is the cheapest. Sometimes, instead of cutting out a piece of Bark, and leaving the wound open, two horizontal gashes are made, one a little above the other; in the upper of which is inserted the stalk of a maple leaf, the point of the leaf being inserted in the lower gash, so as to form a sort of cup to receive the manna, and to preserve it from dust and other impurities. The greater part of the manna of commerce is procured in the latter manner, and is imported in chests, in long pieces, or granulated fragments, of a whitish or pale yellow colour, and in some degree transparent. The inferior kind is of a dark brown colour, in adhesive masses, and is moist and unctuous when felt. Manna from the Ash has a peculiar odour, and a sweetish taste, accompanied with a slight degree of bitterness. "It was formerly used in medicine, but is now chiefly used to disguise other drugs in administering them to children."—*Loudon.*

In the deserts of Syria and Arabia is abundantly produced a leguminous plant,* called by Hasselquist, *Ononis Spinosa;* by Tournefort, *Alhagi Maurorum,* from which is procured a substance, answering yet more closely than the above to the description of the manna of the Israelites. At first it resembles drops of honey, but candies when exposed to the air, granulating into pieces not larger than coriander seed. The Arabs have a tradition that it fell in unusual quantities to supply the Israelites with food during their wandering in the wilderness. The supply of manna on that occasion, however, being purely miraculous, can be explained neither by reference to ancient legends nor modern science. The very meaning of the name, "*what is it,*" would imply that it was a substance with which the Israelites were unacquainted: and the silence of Moses on the occasion is sufficient evidence that he had met with nothing of the kind during his previous residence of forty years in Arabia.

As a timber-tree the Ash is exceedingly valuable, not only on account of the quickness of its growth,

* Leguminous plants are those which produce their seed in pods, *e.g.* the bean, pea, vetch, laburnum.

but for the toughness and elasticity of its wood, in which latter quality it surpasses every European tree. In its younger stages (when it is called *ground-ash,*) it is much used for walking sticks, hoops, and hop-poles; and it matures its wood at so early an age, that an Ash-pole, three inches in diameter, is as valuable and durable for any purpose to which it can be applied, as the timber of the largest tree. "The use of Ash is (next to that of the Oak itself) one of the most universal: It serves the soldier*—and heretofore the scholar, who made use of the inner bark to write on, before the invention of paper. The carpenter, wheel-wright and cart-wright find it excellent for plows, axle-trees, wheel-rings, and harrows; it makes good oars, blocks for pullies, and sheffs, as seamen name them: for drying herrings no wood is like it, and the bark is good for the tanning of nets; and, like the Elm, (for the same property of not being apt to split or scale) is excellent for tenons and mortises; also for the cooper, turner and thatcher; nothing is like it for our garden palisade hedges, hop-yards, poles and spars, handles and stocks for tools, spade-trees, &c. In summer, the husbandman cannot be without the Ash for his carts, ladders, and other tackling, from the pike, spear, and bow, to the plow; for of Ash were they formerly made, and therefore reckoned amongst those woods which, after long tension, has a natural spring, and recovers its position, so as in peace and war it is a wood in highest request: In short, so useful and profitable is this tree next to

* Spears were anciently made of Myrtle, Cornel and Hazel, but Pliny prefers the Ash for that purpose.

the Oak, that every prudent lord of a manor should employ one acre of ground with Ash to every twenty acres of other land, since in as many years it would be more worth than the land itself."—But, we may add, it should be planted in sheltered situations, where the soil is moderately dry. "Some Ash is curiously cambleted and veined, so differently from other timber, that our skilful cabinet-makers prize it equal with Ebony, and give it the name of green Ebony, which their customers pay well for; and when our woodmen light upon it, they may make what money they will of it."—"Lastly, the white and rotten dotard part composes the ground for our gallants' sweet powder; and the truncheons make the third sort of most durable coal, and is, of all others, the sweetest of our forest fuelling, and the fittest for ladies' chambers; it will burn even whilst it is green, and may be reckoned amongst the kinds of wood which burn without smoke."*

Phillips, speaking of the value of Ash-timber, says: "In remote times, when this island was over-run with woods, timber-trees were principally valued for the food which they yielded to herds of swine; and thus, by the laws of Howel Dda, the price of an Ash was rated at 4d., while an Oak or a Beech was put at 120d. At the present time, Ash-timber meets with as ready a sale, and brings nearly as high a price, as the best Oak; and although we do not so frequently meet with large Ash-trees as we do with large Oaks and Elms, yet it will be seen that the natural size of the tree is nearly the same. But as it grows so much more rapidly than the Oak,

* Evelyn's Sylva.

so will it sooner decay than that tree, if not felled at maturity. It is observed, that when the woodpeckers are seen tapping these trees, they ought to be cut, as these birds never make holes in the Ash, until it is on the decay."

This fact has not escaped the notice of the observant Waterton, the well known Ornithologist. "Often," he says, "when arrayed in all the bloom of vegetable beauty, the Ash-tree is seen to send forth from its bole, or from some principal branch, a small fungus, which during the summer increases to a considerable size. It ripens in the autumn and falls to the ground when winter's rain sets in. The bark through which this fungus sprouted is now completely dead, though it still retains its colour; and that part of the wood from which it proceeded is entirely changed in its nature, the whole of its vitiated juices having been expended in forming and nourishing the fungus.

"Nothing remains of its once firm and vigorous texture. It is become what is commonly called touchwood, as soft and frangible as a piece of cork, and, when set on fire, will burn like tinder. In the mean time, the tree shows no sign of sickness; and its annual increase goes on as usual; till, at last, the new swelling wood closes over the part from which the fungus had grown, and all appears to go on right again. But, ere the slow process arrives at this state, the titmouse or the woodpecker will have found an entrance and a place of safety for their incubation. They quickly perforate the distempered bark; and then the tainted wood beneath it yields to their pointed bills, with which they soon effect a spacious cavity. Here then we have the whole mystery unfolded. These birds, which never perforate the live wood, find in this diseased part of the tree, or of the branch, a place suitable to their wants. They make a circular hole, large enough to admit their bodies; and then they form a cavity within, sufficiently spacious to contain their young. Thus does Nature kindly smooth the way, in order that all her creatures may prosper and be happy. Whenever I see these sylvan carpenters thus employed, I say to them, 'Work on, ye pretty birds; you do no harm in excavating there: I am your friend, and I will tell the owner of the tree that you are not to blame.' But his woodman deserves a severe reprimand. He ought to have cut down the tree in the autumn, after the appearance of the fungus."—*Essays on Natural History.*

The author of "*The Journal of a Naturalist*" strongly deprecates the custom of pollarding trees, and more particularly the Ash. "This system of cutting off the heads of the young trees in the hedge-rows is resorted to by the farmer for the purpose of forcing them, thus deprived of their leaders, to throw out collateral shoots, serving for stakes for fences and for firewood. These purposes are effected; but of all hopes of timber, or profit to the proprietor there is an end. No trees suffer more in this respect than the Ash. Prohibitions against mangling trees are usual in agreements; but, with some exceptions in regard to Oak, little attention seems paid to the covenant, as is obvious on the most cursory view of the country in any direction; whereas the Ash is not a less valuable tree than the Oak, from its thriving more universally in all situations, and becoming saleable in a shorter period. One or two generations must pass before an Oak should be felled; but the Ash becomes useful wood while its more respected companion is but a sapling. Young Ash-trees should therefore be more especially guarded, because they are the most likely to suffer, from their producing the greatest quantity of lop in the shortest time. It is by no means an

POLLARD-ASH.

uncommon thing to observe every Ash-tree reduced to stumps by successive pollardings. I am not so silly as to enlarge upon the beauty of what has been called "picturesque farming;" but when we cast our eyes over the country, and see such rows of dark, club-headed posts, we cannot but remark upon the unsightly character they present, and consider that it is neither laudable to deform our beautiful country by connivance at the practice, nor that it is proper attention to individual profit to allow the continuation of it. The Ash, after this mutilation, in a few years becomes flattened at the summit, moisture lodges in it, and decay commences, the central parts gradually mouldering away, though for many years the sap wood* will throw out vigorous shoots for the hatchet. The caterpillar of the goat moth now too commences its ravages, and the end is not far distant. But the wood of the Ash appears in every stage subject to injury; when in a dry state the weevils mine holes through it; when covered by its bark, it gives harbour to an infinite variety of insects, which are the appointed agents for the removal of the timber: the ashen bar of a stile, or a post, we may generally observe to be regularly scored by rude lines diverging from a central stem, like a trained fruit tree, by the meanderings of a little insect (*ips niger*, &c.), being the passages of the creatures feeding on the wood."

Among the most remarkable Ash-trees of which I can find an account are the following:—

Evelyn mentions "divers in Essex which measured in length 132 feet." Phillips says: "At Doriney, near Clare, in the county of Galway, is an old Ash, that at four feet from the ground measures 42 feet in circumference; at six feet high,

* *Alburnum*, or *sap wood*, comprises the light-coloured, recently formed layers of wood adjoining the bark.

30 feet. The trunk has long been quite hollow, and a little school was kept in it. There were a few branches remaining in 1808, which were fresh and vigorous. Near Kennity Church, in the King's County, is an Ash, the trunk of which is 21 feet 10 inches round, and it is 17 feet high before the branches break out. These are of enormous bulk. When a funeral of the lower class passes by, they lay the corpse down for a few minutes, say a prayer, and then throw a stone to increase the heap, which has been accumulating round the root."

Sir Thomas Dick Lauder mentions an enormous Ash tree near the house of Bonhill, in Dumbartonshire, which at four feet from the ground measures 34 feet. It is hollow, and in the inside a little room has been fitted up, nine feet in diameter and eleven in height. "It is floored and surrounded with a hexagonal bench, on which eighteen people can sit, with a table in the middle; and above the door there are five leaden windows. The whole trunk, which is a vast mass, is thickly covered with fresh vigorous branches."

The Great Ash at Woburn, described and figured by Strutt in his *Sylva Britannica*, is an extraordinary specimen of the size which this tree will attain in favourable situations. It is 90 feet high; and the stem alone is 28 feet. At the ground it is 23½ feet in circumference; 20 at 1 foot and 15 feet 3 inches at 3 feet from the ground. The diameter of its spread is 113 feet, and it contains 872 feet of timber.

In the parish of Cury, about six miles from the Lizard point, in Cornwall, stands a very fine

Ash, known by the name of "Cury Great-Tree." Although in dimensions it does not equal those above mentioned, it is worthy of notice as being the largest Forest-tree in the extreme southern promontory of England, and for the veneration in which it is held by the inhabitants of the surrounding district. It is picturesquely situated in the centre of an area formed by the junction of three roads, and extends its bulky arms over the greater part of the open space, a diameter of about seventy feet. At the surface of the ground it measures 27 feet in circumference; at five feet from the ground, 14 feet. It is hollow internally, the cavity at the ground measuring five feet, but presently decreasing to three feet; at about 14 feet from the ground it sends out six spreading limbs, two of which have long been broken off close to the bole, and all the rest are much battered at the extremities.

Some years ago "Cury Great-Tree" enjoyed considerable notoriety from being frequently selected as a place of rendezvous for smugglers; and not a few stories of lawless violence perpetrated under its spreading branches are current in the neighbourhood. An incident related to me by Captain Head is worthy of recording, as affording a striking contrast to the pacific disposition of the present inhabitants of the surrounding district. "I recollect having heard from my father that when he was a boy of about ten years of age (which must have been in the year 1750—3) a large ship was wrecked between the Lizard and Kynance Cove. At that period the miners and others used to assemble in large bodies, whenever they received tidings of a wreck, and hasten to the spot for plunder. On this occasion the tinners of Wendron, a parish about ten miles off, got the start of the other parts of the mining district, and were the first to lade themselves with the booty, which in those days they considered their lawful prey. The men of Breage and Germoe (parishes yet farther distant) on their arrival at this Great-Tree received the information that they had been forestalled; and knowing that most of the valuables would be secured and appropriated before they could reach the scene of rapine, resolved on waiting at this spot and intercepting the plunderers on their return. It was pretty late in the afternoon when the Wendron men, heavily laden with what they considered their rightful gain, arrived at the ambuscade of their no less unscrupulous waylayers. A furious battle ensued, *might* being on either side the only test of *right*, and several men were killed. But what principally distinguishes this fray from many others of the same character, is that a woman who was interested for one of the parties, having deliberately pulled off one of her stockings and placed a large stone in it, mounted a hedge closely adjoining the scene of conflict, and with this unusual but murderous weapon actually beat out the brains of a man of the adverse faction." Now, happily, a better spirit prevails in Cornwall. In these days, the tourist who, at the close of his day's wandering along the cliffs and craggy rocks of the Lizard, rests himself beneath the shade of the great Ash, will have no difficulty in picturing to himself the deepest horrors of a shipwreck on that iron coast; but will scarcely be brought to believe that so dark a deed as that above described was perpetrated on the spot where he is sitting and almost within the memory of living men.*

I must not omit to mention the parent of all the "Weeping-Ashes," which if not approaching in beauty the normal condition of the tree, are so frequently employed to decorate suburban gardens. This singular tree was discovered about the middle of the last century in a field belonging to the Vicar of Gamlingay, near Wimpole, in Cambridgeshire. It was then a very old tree: some of its progeny have indeed attained the age of sixty years. Grafts (for by means of them only is it propagated, have been carried to France, Germany, and even to America.

A very large specimen of the drooping Ash was in 1828 or 1829 removed from Wilson's Nursery at Derby, to Chatsworth, the seat of the Duke of Devonshire, a distance of 28 miles. It was then 50 years old, and weighed nearly 8 tons. By the aid of a machine constructed for the express purpose, and the united efforts of men and horses, its transit was with difficulty effected, though not before the turnpike gates along the line of road had been removed, and a breach made in the park-wall to admit it. In the first year after its trans-

* In confirmation of the statement here made, that the name and occupation of "Cornish wreckers" have disappeared, I may add that in the past year, 1845, a French ship was driven ashore within 12 miles of the scene of the shipwreck alluded to above. The inhabitants of the neighbouring village, Porthleven, rendered every assistance in their power to the unfortunate crew, all of whom were saved. The vessel was driven so high on the shore that she could not be got off and was of necessity sold as she stood on the beach. Greatly to their honour, the poor fishermen, who had bestowed much of their time and labour on the preservation of the vessel, declined any remuneration, and, with one solitary exception, begged that the salvage money might be paid over to the houseless strangers, which was accordingly done.

portation it sent out shoots 12 inches long. A curious weeping Ash is to be seen in the garden of the Vernon Arms, New Road, London. It is trained by trellis work, at a height of 7 feet from the ground over 14 tables and 28 benches, covering a space 36 feet long by 21 feet wide.

Another variety of the Ash (Fraxinus heterophylla) is occasionally to be met with, bearing simple leaves, but is only remarkable for wanting the graceful lightness of foliage which characterizes the common Ash. This, too, is increased by grafting, but is altogether unworthy of being encouraged. Some botanists consider this, but without sufficient grounds, to be a distinct species.

Jesse, in his interesting *Gleanings in Natural History*, gives the following remarkable instance of an extraneous substance being found imbedded in the solid timber of an Ash: "A person on whose accuracy and veracity I can place every reliance, informed me that hearing from some of his brother-workmen, that in sawing up the butt of a large Ash-tree, they had found a bird's nest in the middle of it; he immediately went to the spot, and found an Ash cut in two longitudinally on the saw-pit, and the bird's nest nearly in the centre of the tree. The nest was about two-thirds of a hollow globe, and composed of moss, hair and feathers, all seemingly in a fresh state. There were three eggs in it, nearly white and somewhat speckled. On examining the tree most minutely with several other workmen, no mark or protuberance was found to indicate the least injury. The bark was perfectly smooth and the tree quite sound." In endeavouring to account for this curious fact, we can only suppose that some accidental hole was made in the tree before it arrived at any great size, in which a bird had built its nest, and forsaken it after she had laid three eggs. As the tree grew larger, the bark would grow over the hole, and in process of time the nest would become imbedded in the tree.

I cannot better conclude this chapter than by quoting Strutt's description of the Ash at Carnock in Stirlingshire, with which he closes his *Sylva Britannica.* "Its dimensions are as follows: 90 feet in height; 31 feet in circumference at the ground; 19 feet 3 inches at five feet from the ground, and 21½ feet at four feet higher up. The solid contents of the tree are 679 cubic feet. It was planted about the year 1596, by Sir Thos. Nicholson of Carnock, Lord Advocate of Scotland in the reign of James VI. It is at present in full vigour and beauty, combining airy grace in the lightness of its foliage and the playful ramifications of its smaller branches, with solidity and strength in its silvery stem and principal arms. Delightful indeed is it to contemplate the variety and surpassing beauty of many of these "houses not built with hands," proclaiming to the viewless winds, to the eyes of Heaven and to the heart of man the wisdom and the love of the Eternal Architect, whose fiat calls them into existence, and whose benevolence wills them to live."

THE BOX.

BUXUS SEMPERVIRENS.

Natural Order—EUPHORBIACEÆ.

Class—MONŒCIA. *Order*—TETRANDRIA.

MANY of my readers probably are acquainted with the subject of the present chapter only as a neat edging for flower-beds, or as a shapely bush in the formal garden of some antiquated manor-house: yet the Box-tree has a very good claim to be considered a native British tree. Its right is certainly disputed both by some of the old botanists, and by the more recent authors who quote their opinions; but inasmuch as it is in undeniable possession of at least one extensive district in England, and has been so long enough to give to that one the name of Box-hill, I think we are quite justified in advocating its claims to be considered a native tree. Besides this, not only did it give name to Boxley in Kent, and Boxwell in Gloucestershire, which would prove, at least, that it has grown at these places from time immemorial, but it is expressly mentioned by several authors as a native. Gerard, for instance, who wrote in Elizabeth's reign, says: "It groweth upon sundry waste and barren hils in Englande." Evelyn says: "These trees rise naturally at Boxley in Kent, and in the county of

Surry, giving name to that chalky hill* (near the famous Mole or Swallow) whither the ladies and gentlemen, and other water-drinkers from the neighbouring Ebesham Spaw, often resort during the heat of summer to walk, collation, and divert themselves in those antilex natural alleys and shady recesses among the Box-trees, without taking any such offence at the smell which has of late banished it from our groves and gardens." Gilpin, too, is of the same opinion; speaking of Box-hill, he says: "This plant grows here in full luxuriance, in its native uncultivated state, marking the road on the right with great beauty." This is, I believe, the only place in Great Britain in which the Box grows in profusion in its wild state. Here it attains the height of about fifteen or sixteen feet, and gives to the scenery quite a foreign character, the mellow tint of its foliage harmonizing well both with the grey of its stem and the richer green of any other tree which may happen to be associated in the landscape with it; and at seasons when other trees are out of leaf it displays an unconsciousness of winter, which no artificial shrubbery can compete with.

Evelyn says, quaintly, but with great propriety: "He that in winter should behold some of our

* Boxhill. The Hon. Daines Barrington, in a paper inserted in the Philosophical Transactions for 1769, says: "Now we happen to know that this hill was so called from an Earl of Arundel's" (the famous antiquary) "having introduced this tree in the reign of James or Charles the First." Barrington does not state whence he obtained his knowledge, nor does he account for the fact that a naturalist of the preceding century found it growing on "the waste and barren hils in Englande," at least forty years before James the First came to the throne.

highest hills in Surry clad with whole woods of these trees, for divers miles in circuit (as in those delicious groves of them belonging to the Honourable, my Noble Friend, the late Sir Adam Brown, of Beckworth Castle, might, without the least violence to his imagination, fancy himself transported into some new or enchanted country; for, if in any spot of England,

> "Hic ver assiduum, atque alienis mensibus æstas,
> 'tis here
> Eternal spring and summer all the year."

Most other shrubs, if left to themselves, in a few years outgrow their beauty, becoming bare near the ground, and assuming an unsightly, straggling appearance. But the Box retains its shape for many years, and, as it here forms a thick and extensive coppice, it gives to the country a character as pleasing as it is unusual.

In the East it attains a much larger size than with us, and is mentioned in the Sacred volume in conjunction with several of the largest forest-trees: "I will set in the desert the fir-tree, and the pine, and the box-tree together: that they may see, and know, and consider, and understand together, that the hand of the Lord hath done this, and the Holy One of Israel hath created it." (*Isaiah* xli. 19, 20.)

As a cultivated tree it was formerly much valued by practitioners of the topiary art,* for which

* Topiary work, or, *the art of cutting the Box and other trees into artificial forms*, was carried to such an extent among the Romans, that both Pliny and Vitruvius use the word *topiarius* to denote the art of the gardener: a proof that, as far as ornament was concerned, the art of clipping was considered the highest accomplishment that could be

it is better adapted than any other tree, owing to the closeness of its habit of growth and its suffering no injury from the frequent use of the shears. It is a slow grower, attains a great age, and will thrive in most soils, and at almost any temperature. It was so trained as to represent architectural devices, figures of men and animals, arcades, and various other forms. The method adopted in order to produce these various semblances was to enclose the tree in a light frame of wickerwork, constructed in the shape required, and to cut back the shoots which protruded till a solid mass of verdure was produced. The wickerwork was then removed, and the Box-tree compelled to retain its grotesque shape by frequent use of the shears or knife. Even now we may occasionally fall in with a vegetable globe or some other such absurdity: but gardeners now-a-days, instead of wasting their time in distorting Nature, employ it more profitably in assisting her to produce new varieties, or studying how to rear and acclimatize new species, of useful and ornamental plants.

Various extracts and perfumes were formerly made from the leaves and bark of this tree, and were considered specifics for a yet greater variety of diseases. Modern science has, however, dis-

possessed by a gardener among the ancient Romans. This appears to have been equally the case in Europe in modern times; gardeners, even so late as the time of the Commonwealth, being called by Commenius "pleachers" (from the old word *pleach* "to interweave.") About the middle of the seventeenth century, the taste for verdant sculpture was at its height in England; and, about the beginning of the eighteenth, it afforded a subject for raillery for the wits of the day, soon afterwards beginning to decline."—(Loudon.)

carded them all. There seems yet to remain a lingering belief that a decoction of the leaves strengthens the hair; but in by-gone days its efficacy was deemed greater even than that of any of the modern nostrums recommended for the same purpose.

Old Gerard, who was sufficiently credulous in other and less plausible matters, (for example, that the Barnacle-goose owed its origin to the Oak,) very wisely observes, that the Box "is more fit for dagger-hafts than to make medicines."

Box-wood contains a powerful sudorific principle with a bitter taste, which has been separated and named *Buximia.* M. du Petit Thouars some years since stated to the Philomathic Society of Paris, that more Box-wood than hops entered into the composition of almost all the beer in France. Olivier de Serres recommends the branches and leaves of the Box as by far the best manure for the vine; not only because it is very common in the South of France, but because there is no plant that by its decomposition affords a greater quantity of vegetable mould. Wordsworth relates that "in several parts of the north of England, when a funeral takes place, a basinful of sprigs of Box is placed at the door of the house from which the coffin is taken up; and each person who attends the funeral takes one of these sprigs, and throws it into the grave of the deceased."*

* Twigs of Rosemary were formerly carried, in like manner, by persons attending funerals. In many parts of the Continent the custom still continues. Hogarth, in one of his pictures, represents the mourners carrying small sprigs. In South Wales it is yet common for those who accompany the corpse to carry sprigs of

> "Fresh sprigs of green box-wood, not six months before,
> Filled the funeral basin at Timothy's door."

In the north of Devon newly-made graves may frequently be seen decked with sprigs of Box and other village evergreens: and it takes its place among Holly and Laurel as an ornament of our churches generally, at Christmas.

By the ancients Box-wood was highly valued as a material for musical instruments, *buxus*,* the name by which it was known, often standing for a "flute;" and in our own country it is said by Evelyn to have been "of special use for the turner, engraver, mathematical-instrument maker, comb, and pipe-maker, who give great prices for it by weight, as well as measure; and by the seasoning, and divers manner of cutting, vigorous insolations, politure and grinding the roots of this tree (as of even our common and neglected Thorn), do furnish the inlayer and cabinet-maker with pieces rarely undulated, and full of variety. Also of Box are made wheels or shivers, (as our ship-carpenters call them,) and pins for blocks and pulleys; pegs for musical instruments: nut-crackers, weavers' shuttles, hollar-sticks, bump-sticks, and dressers for the shoemaker, rulers, rolling-pins, pestles, mall-balls, beetles, tops, chess-men, screws, bobbins for bone-lace, spoons, nay the stoutest axle-trees."

"The Box-wood used by the cabinet-makers

Rosemary, or Yew, which they strew on the coffin after it is lowered into the grave.

* Buxus was also used to signify "a comb" and "a boy's top," which were usually made of the same material.

and turners in France is chiefly that of the root. The town of St. Claude, near which is one of the largest natural Box-woods in Europe, is almost entirely inhabited by turners, who make snuff-boxes, rosary beads, forks, spoons, buttons, and numerous other articles. The wood of some roots is more beautifully marked, or veined, than that of others, and the articles manufactured vary in price accordingly. The wood of the trunk is rarely found of sufficient size for blocks in France; and when it is, it is so dear, that the entire trunk of a tree is seldom sold at once, but a few feet are disposed of at a time, which are cut off the living tree as they are wanted. Boxes, &c., formed of the trunk are easily distinguished from those made of the root, the former always displaying a beautiful and very regular star, which is never the case with the latter." (*Loudon.*)

Box is the hardest and heaviest of all European woods, the only one among them that will sink in water, or that is sold by weight. By far the most important use to which Box-wood is applied is as a material for wood-engraving, an art which has now attained such perfection, and is in such great request for the illustration of books, that it may not be uninteresting if I here introduce a short sketch of its history.

A method of multiplying copies of a pattern by means of a stamp was known to the ancient Babylonians, as may be proved by an examination of some bricks brought from the site of the city of Babylon, and preserved in the British Museum. These bear in them characters evidently produced by pressure from a wooden block while the clay was in a soft state. At a later period, the Chinese and Indians were accustomed to print on paper, cotton, and silk, (though it does not appear that they had carried the art to such perfection as to delineate figures,) long before the custom was practised in Europe. In the thirteenth and fourteenth centuries, when writing was an accomplishment confined to the learned, a wooden stamp was used in the place of a sign-manual for attesting written documents; and in the fifteenth century, or perhaps earlier, the art was applied to stamping figures on playing cards. If the earliest cards bore designs at all resembling the grotesque figures on modern specimens, wood-engraving was as yet very far from having any pretension to be considered one of the fine arts, or in the least degree connected with them. Most probably they are exact copies, for, so utterly unnatural are the kings and queens depicted on them, that it is scarcely possible they can be anything else than traditional absurdities.* A modern playing-card may therefore be considered as affording a fair specimen of the perfection of wood-engraving in the fifteenth century. The next step in advance was the delineation of figures of the Saints, on which account the art received the patronage of the Church. The oldest wood-cut of which there is any authentic record, is one of St. Christopher carrying an infant Saviour through the water, and bearing the date

* A similar instance of obstinate adherence to an old, and therefore familiar pattern, a long way behind the existing state of the Arts, may be observed in the never-ending "willow pattern" on earthenware.

of 1423. It is of folio size, and coloured in the manner of our playing cards.

Cristofori faciem die quacunque tueris
Illa nempe die morte mala non morieris.* **Millesimo CCCC° XX° terno.**

* *Translation*:—Gaze on the face of Christopher every day,
So on that day thou shalt not die an evil death. 1423.

Such engravings appear to have been distributed as devotional pictures among the laity,

and to have been occasionally preserved by the monks, who pasted them into the earliest printed books with which they were furnished. That of St. Christopher above alluded to, was discovered in the monastery of Buxheim, near Meiningen, and is now in the possession of Earl Spencer. Collections of them appear also to have been published before the invention of printing from moveable types, for the use of those who either were unable to read, or could not afford to purchase a manuscript copy of the Scriptures. The most important of these is the *Biblia Pauperum*, or Poor Preacher's Bible, a collection of historical subjects from the Old and New Testament, accompanied by explanatory inscriptions in Latin. This appears to have been a most popular book, for not only are there many copies of it, struck from different blocks, but it was also repeatedly printed long after the introduction of printing with moveable types. Another work of the same kind, *The Apocalypse, or History of St. John*, was published about 1434. Of this there are six different editions, and the execution of some of the wood-engravings evinces considerable ability.

The history of the art here divides into two branches, with one of which, the art of printing, properly so called, I must leave my readers to acquaint themselves from other sources. In the fifteenth century we find the two combined in the *Psalter* published by Faust and Schæffer at Muntz. The initial letters, engraved in wood, are executed in the most beautiful style of the art. This custom soon became general, and was introduced into England by Caxton, in 1476. Not long after this, Mair in Germany published prints, the dark parts of which were produced by an impression from a copper-plate engraving, the lighter from a wooden block, but of course by two distinct operations. About the same time, Carpi, in Italy, produced wood-cuts by the tedious process of printing on the same paper from three several blocks, the first containing the outline, the second the dark shadows, the third the light tints. But a much greater improvement was effected by Albert Durer, who, by a simpler process, produced wood-cuts in which the figures were more skilfully designed and grouped, the laws of perspective more carefully attended to, and a variety of minor details introduced, which gave to the subject more of the stamp of truth and Nature. The names of various other artists might be mentioned, who from time to time distinguished themselves by the eminence which they attained, until the close of the seventeenth century, when the custom of illustrating books with copper-plate engravings came into vogue, and wood-engraving was entirely neglected, so far as it regarded the delineation of subjects of interest, being employed solely for common decoration. That this should have happened is very remarkable, inasmuch as the superiority of wooden blocks over copper-plates in illustrating printed books is very great. In copper-plate engraving, the lines from which the design is transferred are sunk into the metal, either by the corroding effects of a mineral acid, or by a sharp pointed steel instrument. Consequently the sunken lines must be filled with ink before an impression can be struck off: but in ordinary letter-press printing, a raised surface alone receives the ink and transfers the copy. Hence arises an impossibility of printing both by the same process. But in wood-engraving the thickness being carefully regulated by the height of the type with which it is to be used, it is set up in the same page with the types;

and only one impression is required to print the letter-press and the cut which is to illustrate it. Added to this, the friction (though produced simply by the soft fleshy ball of the thumb) which is required to charge the lines of a copper-plate engraving with ink, soon wears away the sharpness of the lines, and renders every new impression less perfect than its predecessor. But in printing wood-cuts, the whole of the pressure being vertical, there is no perceptible wearing away of the block, so that the goodness of the impression depends only on the materials employed, and the care of the printer.* But even on the supposition that the mechanical advantages of each were equal, the preference must be awarded to wood-cuts for the illustration of printed books, inasmuch as there is a harmony produced in the page by the engraving and letter-press being of the same colour, which is very seldom the case when copper-plate vignettes are introduced with letter-press.

In spite however of all these advantages, the art of engraving on wood declined, and was all but lost, when it was revived in England by the celebrated Thomas Bewick, an artist who not only restored the taste for the art, but executed, in the course of a long and industriously spent life, numerous works, which his most zealous

* In an interesting Memoir of Bewick, prefixed to the sixth volume of Jardine's Naturalist's Library, it is stated, that, "many of Bewick's blocks have printed upwards of 300,000: the head-piece of the Newcastle Courant above 1,000,000; and a small vignette for a capital letter in the Newcastle Chronicle, during a period of twenty years, at least 2,000,000."

followers can scarcely do more than hope to equal. His excellence did not consist in the mere mechanical skill which he displayed: that, great as it was, resulted from his intense desire to embody his exquisitely acute perceptions of Nature—his wood-cuts, therefore, are not simply representations of birds and beasts, just so far like the originals as to enable another person to discover what is meant:—but indexes of his mind, like the solemn sound of Handel's music, the majestic flow of Milton's poetry, the comprehensive exactness of Linnæus's descriptions. No one can have failed to notice this, who has turned over the pages of "The general History of Quadrupeds" or of "British Birds": Nature seems to be alive in all of them; the very tail-pieces, trifling though the subjects of many of them may be, are replete with interest, owing to the remarkable power which the author possessed of catching and pourtraying the peculiar characteristics of Nature, whether animate or inanimate. Much of this taste and skill Bewick imparted to his pupils,* and to the same qualities the modern school of wood-engraving is indebted for its principal excellence.

Several mechanical improvements have of late years been made in wood-engraving and print-

* I was much interested, some years since, in the north of Devon, by falling in with a rustic well, surmounted by a rude stone cross with a wicket gate by its side. It was just the sort of subject that Bewick would have chosen for a vignette. I afterwards found that the proprietor (who I am sure will forgive me for this mention of him) was formerly a pupil of Bewick, and, before his accession to fortune, gave no slight promise of sharing his master's fame. *(See vignette at the end of this chapter.)*

ing; but, however the father of the modern art may be surpassed in skill, it is next to impossible for any one to excel him in excellence of design.

Owing to the numerous illustrated works now almost daily issuing from the press, the number of artists in this line has greatly augmented, and Box-wood has proportionately increased in price.

In 1815, the trees which were cut down on Box-hill produced upwards of £10,000. A great deal of that imported from Turkey, Odessa, and other places is inapplicable to the purposes of the wood-engraver; nevertheless, in London alone, as much is annually consumed in works of art as amounts to many thousands of pounds.

There are, besides the Tree-Box, two varieties of Dwarf-Box, which were formerly much employed in forming patterns in flower-gardens, imitating the designs of embroidery. This fashion is now quite gone out, like topiary work, having given place to the much more rational taste of cultivating various exotic plants; but representations of quaintly figured gardens may yet be seen in old engravings. Dwarf-box is now only planted as an edging to garden-beds, for which its low wiry habit well adapts it, preventing the loose earth from falling into the path, without rising high enough to shade the plants in its neighbourhood, or affording a secure refuge for vermin. It may be propagated by dividing the roots, or by planting cuttings in autumn. The best time for clipping Box is in June, when the new shoots soon obliterate all traces of the shears.

The flower of the Box is inconspicuous, being of a greenish yellow colour, and growing in clusters in the axils* of the leaves. It ripens its seed at Box-hill: but flowers have never been observed on the dwarf variety.

* Axil, Latin, *axilla*, the arm-pit ; in botanical phraseology "the angle between the leaf-stalk and stem."

THORN AT NEWHAM.

THE HAWTHORN.

Cratægus Oxyacantha.

Natural order—Rosaceæ.

Class—Icosandria. *Order*—Pentagynia.

There is, I think, no tree, the simple mention of which excites such pleasurable emotions as the Hawthorn. Never attaining a remarkable size, neither stately in growth, nor graceful in form, it yet possesses an interest to which many a loftier and more elegant child of the forest cannot aspire. We may see it applied to the most homely and unromantic purposes, clipped by the hedger's shears of every particle of natural spray, and reduced, as it were by line and plummet, to the uniform proportions of a mere verdant wall; yet the tree to which the mind reverts when the Hawthorn is mentioned is independent of any such associations. It does not, it is true, carry us away to forests or woodland mountains, to the wild fastnesses of Nature, where men and the things of men have no place. Were we acquainted with it only in such situations, it would want half its interest—but it recurs to the memory as the necessary appendage of the village, to which, in our earlier years, it was our highest privilege to make our holiday excursions —the veteran record of our infantile sports, remaining unchanged while the stern realities of life have been working in ourselves a change too perceptible—a common shelter from sun or shower to the rude patriarchs of the hamlet, the same group (nearly, for some are not) that half a century ago, tottered as feebly to their childish amusements, as now they do to their shady seat beneath its branches, and from the self-same cabins too—and the contemporary of all the bygone sports that old and young loved to look back upon, or forward to, with equal interest.

The Hawthorn, too, is a tree which, from its association with the village festivities of the first of May, possesses a kind of antiquarian interest, which is deepened by the recollection that it illustrates "the simple annals of the poor." The first day of the month, from which it derived its name, "May-bush," was formerly a general rustic holiday, looked forward to, and prepared for, with as much zest as accompanies many a nobler entertainment; and it was a matter of no little solicitude whether the Hawthorn would be fully blown in good time; for a "bunch of May" was the crowning ornament of the May-pole, and encircled the head of the May-queen, her consort for the day being crowned with the more manly Oak.

Before the alteration of the style* in 1752, the Hawthorn rarely failed to be in flower in

* The ancient church calendar was constructed on the erroneous supposition that the year contained 365¼ days exactly, being nearly twelve minutes too much. The error, therefore, in 129 years amounts to a whole day. In consequence of the inconvenience which was found to result from this error during a long course of years, Pope Gregory XIII., in the month of March, 1582, issued a brief, in which he abolished the old calendar, and substituted that which has since been received in all Christian countries, except Russia, under the name of the *Gregorian Calendar, or New Style. (N.S.)*

good time: but since that period, May-day falling eleven or twelve days earlier, its blossoms are rarely fully expanded even in the south of England, until the second week in the month.* In mountainous districts, the Highlands, for instance, it is frequently in full perfection so late as the middle of June.

By the ancient Greeks its flowers were made the emblem of Hope, and it was probably regarded in the same light by the Romans, as we read that its wood was chosen to make the torch carried before the bride at nuptial processions. In some countries it is regarded with a kind of veneration, from being believed to be the tree used to form the crown placed on our Blessed Saviour's head before His Crucifixion. Whether or not this opinion be a correct one is scarcely a fit subject for discussion in this or any other work. But if it really be the case, it is not improbable that it was selected by the Roman soldiers with the

Gregory, in order to restore the commencement of the year to the same place in the seasons that it had occupied at the time of the Council of Nice, (A.D. 325,) directed the day following the feast of St. Francis, that is to say the 5th of October, 1582, to be reckoned as the 15th of that month. The New Style was adopted in Britain in 1752; from that year till 1800, May-day fell eleven days earlier; and during the present century it falls twelve days earlier than when calculated by the Old Style; *(O.S.)* May-day of the *Season*, being now the 13th day of the month.

* I have, however, seen it in Devonshire so early as the 29th of April; and in the present year, 1846, it was gathered in Cornwall on the 18th of April. So unusually mild was the season this year, that the Oaks at Clowance, Cornwall, had made shoots between two and three inches long on the 11th of April; though it not unfrequently happens that the Oak is not sufficiently in leaf "to hide King Charles" on the 29th of May. The blossom of the Hawthorn, though early, was so exceedingly scarce this year, that many trees might be searched in vain for a single sprig, and scarcely one tree in a hundred bore an average crop of flowers.

object of making the emblem of hope and happiness the instrument of inflicting pain. Such a motive would accord well with the spirit which demanded the Cross and the purple robe. In some parts of France, the country people affirm that the Hawthorn utters groans and sighs on the evening of Good Friday; and when a thunder-storm is impending they gravely adorn their hats with a bunch of its leaves, in the belief that, thus protected, the lightning cannot touch them. It is also related, that on the morning which followed the horrible massacre of the French Protestants by the Roman Catholics on St. Bartholomew's day, a Hawthorn in the churchyard of St. Innocent, in Paris, suddenly put forth its blossoms for the second time.

A custom exists at the Seven Churches, Glendalough, in the County of Wicklow, of hanging shreds of clothing to an old Thorn which overshadows a Holy Well, on the day on which the Patron Saint Kevin is commemorated. The same practice is said to be common in many other parts of Ireland. It is hard to say in what superstitious belief this singular custom originated, or what benefit the deluded fanatics suppose to accrue to themselves from its observance. It is not, however, confined to the Roman Catholics of Ireland, for the Mahommedans of Africa, and the Pagans of South America, practise a similar observance. "Suez," says a traveller in the East, "was distant twenty-four miles, and these were accomplished in four hours and a half. Only two small trees are to be met with in the desert—a space of eighty-four miles—one of which is decorated with, and consecrated to, the rags of the pious pilgrims who cross the sandy and rocky waste over which we passed; *they en route* to Mecca, *we* to a less holy shrine. The tree is thickly covered with pendent fragments of the well-worn clothing of countless pilgrims, deposited there in memory of their desert journey."* Darwin, in his "*Journal of Researches into the Natural History of the Countries visited during the Voyage of H. M. S. Beagle round the World*," in the account of his journey from the mouth of the Rio Negro to Buenos Ayres, says: "Shortly after passing the first spring, we came in sight of a famous tree, which the Indians reverence as the altar of Walleechu. It is situated on a high part of the plain, and hence is a landmark visible at a great distance. As soon as a tribe of Indians come in sight of it, they offer their adorations by loud shouts. The tree itself is low, much branched, and *thorny:* just above the root it has a diameter of about three feet. It stands by itself without any neighbour, and was indeed the first tree we saw; afterwards we met with a few others of the same kind, but they were far from common. Being winter the tree had no leaves, but, in their place, numberless threads, by which the various offerings, such as cigars, bread, meat, pieces of cloth, &c. had been suspended. Poor Indians, not having anything better, only pull a thread out of their ponchos, and fasten it to the tree. Richer Indians are accustomed to pour spirits and maté into a certain hole, and likewise to smoke upwards, thinking thus to afford all possible gratification to Walleechu. To complete the scene, the tree was surrounded by the bleached bones of horses which had been slaughtered as

* "*A Year and a Day in the East.*"

sacrifices. All Indians of every age and sex make their offerings: they then think that their horses will not tire, and that they themselves shall be prosperous. The Gaucho (countryman) who told me this, said that in time of peace he had witnessed this scene, and that he and others used to wait till the Indians had passed by for the sake of stealing from Walleechu the offerings. The Gauchos think that the Indians consider the tree as the god itself; but it seems far more probable that they regard it as the altar."

Neither of the authors quoted above states to what species the tree belongs. Darwin, however, mentions that the one seen by him was "thorny," and if, as is most probable, the African tree was an Acacia, which is copiously furnished with thorns, the points of resemblance are very striking and remarkable. The latter tree may have approached even more closely in appearance to our Hawthorn. Bruce says that the Arabs regard with particular reverence a shrub or tree which is very like our Hawthorn both in form and flower. It was with a branch of this tree, which they call "El-vah," that they believe Moses to have sweetened the waters of Marah. Trees in the desert, I need scarcely remark, on the rare occasions when they are found, are always in the vicinity of wells.

It is far from improbable that the legend of "The Glastonbury Thorn" was originally connected with some superstitious veneration of the Hawthorn, yet more ancient than itself. According to this legend, Joseph of Arimathæa, attended by twelve companions, came to preach the Gospel in Britain, and landed on the Isle of

Avelon.* Here he fixed his staff in the ground (a dry Thorn sapling, which had been his companion through all the countries he had passed through) and fell asleep. When he awoke, he found, to his great surprise, that his staff had taken root, and was covered with white blossoms. From this miracle, he drew a very natural conclusion, that as the use of his staff was taken from him, it was ordained that he should fix his abode in this place. Here, therefore, he built a chapel, which, by the piety of succeeding times, increased to its subsequent magnificence. Gilpin, in his *Observations on the Western Parts of England*, gives the following amusing account of the veneration with which it was regarded at no more distant period than the close of the last century. "I should ill deserve the favours I met with from the learned antiquarian who has the care of these ruins, though he occupies only the humble craft of a shoemaker, if I did not attempt to do some justice to his zeal and piety. No picturesque eye could more admire these venerable remains for their beauty, than he did for their sanctity. Every stone was the object of his devotion. But above all the appendages of Glastonbury, he reverenced most the famous Thorn which sprang from St. Joseph's staff, and blossoms at Christmas.

"It was at that time," he said, "when the King resolved to alter the common course of the year,† that he first felt distress for the honour of the house of Glastonbury. If the time of Christmas

* The high ground on which the Abbey of Glastonbury stands is thus named, and tradition asserts that it was in remote times really an island, the meadows around it having been since formed by the retiring of the sea.

† The alteration of the Calendar alluded to at p. 180.

were changed, who could tell how the credit of this miraculous plant might be affected? In short, with the fortitude of a Jewish seer, he ventured to expostulate with the King upon the subject; and informed his Majesty, in a letter, of the disgrace that might possibly ensue if he persisted in his design of altering the natural course of the year. But though his conscience urged him upon this bold action, he could not but own that the flesh trembled. He had not the least doubt, he said, but the King would immediately send down and have him hanged. He pointed to the spot where the last Abbot of Glastonbury was executed for not surrendering his Abbey; and he gave us to understand there were men now alive who could suffer death, in a good cause, with equal fortitude. His zeal, however, was not put to this severe trial. The King was more merciful than he expected, for though his Majesty did not follow his advice, it never appeared that he took the least offence at the freedom of his letter."

Both Gilpin and his simple-minded informant were in error in supposing the tree then standing to have been the identical one with which the legend is connected. The original "Holy Thorn," which stood on Weary-all-hill, (the spot where Joseph and his companions are said to have sat down *all-weary* with their journey,) originally had two distinct trunks, one of which was destroyed by a Puritan in Queen Elizabeth's reign, and the other, together with many yet more interesting relics of antiquity, shared the same fate during the Great Rebellion. If we may credit James Howell, the author of "Dodona's Grove," (printed in 1644,) the mistaken fanatic who completed the work of destruction did not go unpunished: "And he was well serv'd for his *blind* Zeale, who going to cut doune an ancient white *Hauthorne-Tree*, which, because she *budded* before others, might be an occasion of *Superstition*, had some of the *prickles* flew in to his eye, and made him Monocular."*

There are, however, still in existence two trees of the same description, evidently much above a

* In Ireland, to the present day, it is the popular belief that "no one will thrive after rooting up an old Thorn." Some years since a gentleman residing in Carrickfergus, co. Antrim, employed as his gardener an old artilleryman, named Peter S***, who had been invalided in consequence of wounds received in battle, and passed among his comrades as a brave soldier. One day Peter received directions to uproot a "reverend Hawthorn," which, together with the hedge in which it stood, was to make way for some improvements in the garden. He immediately set to work, and soon cleared the hedge of all that grew in it except the Thorn, the roots of which had penetrated deeply into the ground, and which remained untouched. Next day, the gentleman asked him why the tree had not been removed as he desired. Peter answered, "that it was hardly possible—that it would be dangerous to attempt it." His master remonstrated with him, explaining why it was necessary that the Thorn should be included in the order for removal, and left him with a strict injunction to set about the task immediately, which he, very reluctantly, then prepared to do. Next day, however, to his surprise, he found the devoted tree still maintaining its ground, erect and uninjured. On sharply questioning the offender why he had not followed his directions, poor Peter, with the utmost solemnity, assured him that "he had commenced the work, but that at the moment his pick-axe struck the root of the tree he received a violent blow from some invisible hand that made him stagger and almost fall to the ground—moreover, that on going home, he found that just at the same hour, and he had no doubt, at the very same instant, his wife had experienced a similar blow." After this his master did not urge him further in the matter, but got some other person to extirpate the mysterious tree, and the task was accomplished without any further evil result. Crofton Croker, who is most learned in the superstitions of Ireland, remarks that, according to the popular belief, "On *May-eve* the evil Elves seem to be particularly active and powerful: to those to whom they are inimical they *give a blow unperceived*, the consequence of which is lameness." There can be little doubt that these two superstitions are connected in their origin with that recorded in the text respecting the Glastonbury Thorn.

hundred years old, which no doubt were either grafts, or sprung from seeds, of the original tree. From one of these, which stands within the precincts of the Abbey, in a garden adjoining St. Joseph's Chapel, I received, on the 11th of February, 1846, a sprig, in full leaf, and furnished with perfectly formed flower-buds. The tree from which it was gathered measures two and a half feet in circumference, and I was assured by the vicar of Glastonbury, Dr. Parfitt, that it had been budding and blossoming since Christmas. It blossoms a second time in May, and from these latter flowers only is fruit produced. Formerly, the blossoms were so highly valued, that they were sold at Bristol, and even exported to various parts of Europe, and it is still propagated by grafts in the gardens of the curious, but only on account of the strange efforts which it annually makes to commence spring in mid-winter.

Miss Strickland, in her *Lives of the Queens of England*, mentions that its branches were deemed worthy of being presented to royalty. "Christmas," says Père Cyprian, "was always observed in this country, especially at the King's palaces, with greater pomp than in any other realm in Europe." Among other ancient ceremonies now forgotten, he mentions a pretty one, in which a branch of the Glastonbury Thorn, which usually flowers on Christmas-eve, used to be brought up in procession, and presented in great pomp to the King and Queen of England on Christmas morning. Père Gamache, in mentioning this ceremony, says, this blossoming Thorn was much venerated by the English, because in their traditions they say that St. Joseph of Arimathæa brought to Glastonbury a thorn out of our Lord's Crown, and planting it in the earth, it burgeoned and blossomed, and yearly produced blossoms to decorate the altar on Christmas-eve mass—

> "That only night in all the year
> Saw the stoled priest the chalice rear."
>
> *Wordsworth.*

The Père seems to enjoy very much the following anecdote of Charles I., though it was against the Catholics:—"Well!" said the King, extending his hand, one Christmas-day, to take the flowering branch of Glastonbury Thorn, "this is a miracle, is it?" "Yes, your Majesty," replied the officer who presented it, "a miracle peculiar to England, and regarded with great veneration by the Catholics here." How so," said the King "when this miracle opposes itself to the Pope?" (every one looked astonished in the royal circle, papist and protestant.) "You bring me this miraculous branch on Christmas-day, old style. Does it always observe the old style, by which we English celebrate the nativity, in its time of flowering?" asked the King. "Always," replied the venerators of the miracle. "Then," said King Charles, "the Pope and your miracle differ not a little, for he always celebrates Christmas-day ten days earlier by the calendar of new style, which has been ordained at Rome by papal orders for nearly a century." This dialogue probably put an end to this old custom, which, setting all idea of miracle aside, was a picturesque one; for a flowering branch on Christmas-day is a pleasing gift, whether in a court or a cottage.

The same authoress thus accounts for the fact that the Hawthorn was selected to be the distinguishing badge of the House of Tudor. After the battle of Bosworth, in which Richard III. was slain on Redmore Heath, and his body ignominiously stripped, "the crown was hidden by a soldier in a Hawthorn bush, but was soon found, and carried back to Lord Stanley, who placed it on the head of his son-in-law, saluting him by the title of Henry VII., while the victorious army sang *Te Deum* on the blood-stained heath. It was in memory of the picturesque fact that the red-berried Hawthorn once sheltered the crown of England, that the House of Tudor assumed the device of a crown in a bush of fruited Hawthorn. The proverb of "Cleave to the crown though it hang on a bush," alludes to the same circumstance."

The sight of the Hawthorn always recalls images of rural life; but we must go back to a somewhat remote period to find it invested with its full honours. During the reign of Henry VIII. May-sports were the favourite diversion of all classes, not only in the country, but even in London. On the eve of May-day the citizens used to go in companies to the neighbouring woods and groves, some to Highgate or Hampstead, some to Greenwich, some to Shooter's Hill; there the night was spent in cutting down green branches, in preparing the May-pole, and in a variety of sports and pastimes. On their return early in the morning, the revellers adorned the May-pole with flowers and foliage from one end to the other, the pole itself being previously painted with the most brilliantly variegated colours. The pole was dragged to its destination by a large number of oxen, each ox having a nosegay of flowers tied to the tips of his horns: men, women, and children, all dressed in their gayest habiliments and laden with green boughs, completed the procession. As they passed through the streets of London, they found

> "Each street a park,
> Made green, and trimm'd with trees;"

the church-porches decorated

> "With Hawthorn buds, and sweet eglantine,
> And garlands of roses;"

they heard music sounding from every quarter, and here and there they beheld in their way some May-pole, preserved from the last year, already elevated, and a wide circle of beaming faces dancing round it. The church of St. Andrew the Apostle was called St. Andrew *Undershaft*, from the circumstance that from time immemorial a May-pole or *shaft* had been set up there, which towered considerably *above* the church-tower. Long streamers or flags were now attached to the pole, which was then finally reared to its proper position, amidst the loud cheers of the multitudes gathered round. Summer-halls, bowers, and arbours were now formed near it; the Lord and Lady of the May were chosen, and decorated with scarfs, ribbons, and other braveries; and then the dances, feastings, and merriment of the day fairly began. The King himself frequently took part in these festivities, for, as we learn from *Hall's Chronicle*, "his Grace being young, and not willing to be idle, rose in the morning very early to fetch May or green boughs, himself fresh and richly apparelled, and clothed all his knights, squires, and gentlemen in white satin, and all his guard and yeomen of the crown in white sarcenet. And so went every man with his bow and arrows shooting to the wood, and so repaired again to the court, every man with a green bough in his cap; and at his returning, many hearing of his going a-Maying were desirous to see him shoot; for at that time his Grace shot as strong and as great a length as any of his guard." During the Great Rebellion, the Parliament ordered that "all and singular May-poles be taken down." When Charles II. ascended the throne, the famous May-pole of the Strand* was restored with great pomp and rejoicing, amidst multitudes of people, whose shouts and acclamations were heard from time to time throughout the whole day. When this pole had ceased to be the centre of the merry May-day circles, and the interest with which it was originally regarded had faded away, it was given to Sir Isaac Newton, and by his directions removed to Wanstead, to support the then largest telescope in the world.†

Of late years the celebrity of the Hawthorn as the symbol of May-day festivities has greatly declined. In London, the number of those,

> "That do the fair and living trees neglect,
> Yet the dead timber prize,"

is so vastly increased, that the May-bush "swells its gems" and "salutes the welcome sun" without exciting a passing thought. The only class who, now-a-days,

> "With due honour usher in the May,"

---

* "Amidst the area wide they took their stand,
Where the tall May-pole once o'erlooked the Strand."
Pope.

† Knight's London, vol i. p. 174.

are the poor chimney-sweeps, who, on this their single holiday, put off their sable suit for one day in the year, to deck themselves with flowers and green branches, and, after all, gain but little sympathy for their "maimed rights." In the rural districts we may see, here and there, the tall May-pole standing all the year round, but never decked with flowers, never made the centre of festivity. In a few remote parishes, the poor farmer's boy yet rises earlier on May-morning than on other days, and hastens to attach a branch of Hawthorn to the cottage doors, claiming as a reward, when the inmates are a-stir, a slice of bread and cream; and, in some few towns and villages, principally in the West of England, children on May-day carry round from door to door, garlands of flowers decorated with birds' eggs, and beg contributions of half-pence. But, as far as regards legends, or the merry days of old, the Hawthorn has fallen into the "sere and yellow leaf."

I am indebted to a friend, the Rev. F. Webber, for the following account of the effort made by the celebrated scholar Dr. Parr to keep up the festivities of May-day. "During one of my short vacations in the year 182—, accompanied by a friend who is now rector of a parish in Dorsetshire, I took a few days' ramble through some of the most interesting parts of the county of Warwick. At Leamington we fell in with a young gentleman, who, after introducing us to the localities most worthy of note in the neighbourhood, added to the obligation we were already under to him for his courtesy, by enabling us to become acquainted with the famous Dr. Parr, who, he told us, added to his ardent love of ancient lore, a scarcely less anxiety to preserve in their pristine purity all the festivities of 'The May.'

"Accordingly we proceeded to Hatton, the doctor's residence, and on arriving at the village, we found a gigantic May-pole erected in an open

MAY-POLE.

space, decorated with innumerable flowers, and surrounded by the villagers in their holiday attire. Among a crowd of ladies and gentlemen, we observed a portly personage, attired in full canonicals, and wearing a wig of most orthodox dimensions, whom we could not for a moment hesitate in pronouncing to be the mighty Grecian. On our introduction to him by our friend, he received us with the greatest urbanity and kindness, and immediately allotted us our partners in the dance. I forget the person selected by the doctor to lead off the dance with him, but I think it was the oldest lady of the village. After we had danced for some time, we adjourned by particular invitation to the parsonage, where we were hospitably refreshed after our exertions, the party on the green, I doubt not, being not a little glad to be relieved from the restraint caused by our presence."*

In spite, however, of the exertions made by Dr. Parr and many others, little more than the name of "May-day" remains, and the legendary interest which once attached to the Hawthorn has faded in like manner. Yet, after all, perhaps, we ought not to regret this; for the religious legends afforded, at the best, an unprofitable subject for speculation, and tended rather to lead away the mind from the Creator to the creature, than to stimulate true piety: and the morning's merriment on May-day was but too frequently the forerunner of rioting and dissipation in the evening. In the minds of those who look aright on the works of nature, more real devotion and a greater amount of pleasurable feeling will be excited by the fragrance, symmetry, colouring, and freshness of a Hawthorn wreath, than could be produced by the most plausible monkish tradi-

* The doctor is said to have kept the large crown of the May-pole in a closet of his house, from whence it was produced every May-day, with fresh flowers and streamers, preparatory to its elevation, and to the doctor's appearance in the ring.

tion, or the merriest rustic dance. Let us hope that the love of flowers, now so widely diffused, is based on wiser and better motives. The poet who sang as follows, no doubt sang too truly of the indifference of mankind to the religious impressions which the wonders of the vegetable world are so well calculated to convey. Let us strive to render our hearts more susceptible.

"Where does the wisdom and the power Divine
In a more bright and sweet reflection shine?
Where do we finer strokes and colours see
Of the Creator's real poetry,
Than when we with attention look
Upon the third day's volume of the book?
If we could open and intend our eye,
We all, like Moses, should espy
Ev'n in a bush the radiant Deity.
But we despise these his inferior ways
(Though no less full of miracle and praise):
Upon the flowers of heaven we gaze;
The stars of earth no wonder in us raise,
Though these perhaps do, more than they,
The life of mankind sway." *Cowley.*

Stript, as we have seen, of legendary interest as the Hawthorn now is, and deprived of its high privilege of crowning the Queen of May, it is, nevertheless, still a favourite with all. Not, as I have before said, that it has great pretensions to elegance of form or picturesque beauty; but it possesses qualities which, I may almost say, engage our affections. It is the tutelary guardian of our fields, our orchards, and our gardens; and seems to thrive best, and loves to grow near the rural habitations of men. When the cottager sets about enclosing his bit of garden-ground, the Hawthorn is ready to crown his lowly fence with its protecting and closely woven boughs, which, with their thickset prickles, form an almost impenetrable barrier round the little domain. When arrived at maturity, its stoutest branches are often hacked unmercifully, nearly through their whole dimensions, and forcibly fixed in a direction contrary to their natural growth; yet the lacerated limbs, regardless of this rude treatment, send forth their shoots as vigorously as ever, and accommodate themselves to the humour or convenience of the planter, with all the fidelity of a spaniel. The Hawthorn may be considered, indeed, a domesticated tree, that readily adapts itself to the wishes and wants of man, requiring little care or attention during any period of its growth. Nor are these all its services; every plant that grows near it seems to acquire increased vigour from its friendly shelter and vicinity. The snowdrop, fearless of the tempest, displays its earliest flowers amid the thick covert of the Hawthorn; while the primrose, the violet, and the speedwell are generally its beautiful associates.

Deprived of its Hawthorn hedges, our rural scenery would lose one of its most interesting features, and present to the eye of the painter and the poet little more than a tame and monotonous expanse of country. Not only do they agreeably diversify our immediate vicinities, but when blended by distance give a rich and unrivalled charm to English landscape.

The Hawthorn is also one of the earliest harbingers of summer. What can surpass the beautiful and delicate green of its first unfolding leaves? After surveying from our windows the monotonous and dingy prospect of a long succession of house-tops and chimneys, how refreshing is it to turn our eyes to the green symbol of spring, which tells us that Nature, in her own lovely domain, is quietly preparing her robe of summer beauty! In the balmy month of May, the Hawthorn has no rival. It may then be said to live in an atmosphere of its own fragrance, the whole country being filled with its delicious odour. It has never been my lot to scent the aromatic breezes which are said to float through the air for a distance of many miles from the shores of Ceylon; but I can scarcely think that they are more grateful in themselves, or connected with more delightful associations, than the Hawthorn perfume of an English spring, or, I may add, the summer perfume of an English hay-field. And as to its wreaths of snowy blossoms, I know nothing more beautiful—some with their blossoms fully expanded, dotted with their delicate pink stamens—others, as yet unfolded, resembling little globes of silver set in pedestals of emerald. India may boast of more gorgeous flowers, but surely of nothing more elegant and graceful.

"When first the tender blades of grass appear,
And buds, that yet the blast of Eurus fear,
Stand at the door of life, and doubt to clothe the year,
Till gentle heat, and soft repeated rains,
Make the green blood to dance within their veins:
Then, at their call, emboldened out they come,
And swell the gems, and burst their narrow room;
Broader and broader yet, their leaves display,
Salute the welcome sun, and entertain the day.
Then from their breathing souls the sweets repair
To scent the skies, and purge th' unwholesome air:
Joy spreads the heart, and, with a general song,
Spring issues out, and leads the jolly months along."

Dryden.

In spring and summer the Hawthorn breathes the very soul of rustic poety; its rich profusion of crimson berries contributes largely to the glorious colouring of autumn, and scarcely less to relieve the dreary sameness of winter.

Gilpin, I regret to add, scarcely allows the Hawthorn any claim to be considered an ornament to the landscape. "The Hawthorn," he says, "should not entirely be passed over amidst the minuter plants of the forest, though it has little claim to picturesque beauty. In song, indeed, the shepherd may with propriety,

'tell his tale
Under the Hawthorn in the dale:'

but when the scenes of nature are presented to the eye, it is but a poor appendage. Its shape is bad; it does not taper and point like the Holly, but is rather a matted, round, heavy bush. Its fragrance, indeed, is great; but its bloom, which is the source of that fragrance, is spread over it in too much profusion. It becomes a mere white sheet—a bright spot, which is seldom found in harmony with the objects around it. In autumn the Hawthorn makes its best appearance. Its glowing berries produce a rich tint, which often adds great beauty to the corner of a wood, or the side of some crowded clump."

Now, although the author in this passage professedly speaks of the Hawthorn as forming a part of a "scene of nature presented to the eye," it is clear that he is in reality thinking of it as an ingredient in a painted landscape, and here, it must be confessed, it is not entitled to a prominent place when in full bloom. No painter would admit a mass of glaring white—

whether a patch of snow, a white-washed cottage, or a bush covered with blossom—into the foreground of a picture: either of these would be unsightly in itself, and would draw the eye of the spectator too much away from the more important features of the landscape; but in *living* nature it rarely produces this effect. Sir T. D. Lauder, who dissents from the opinion expressed by Gilpin, thinks that "the Hawthorn,

COMMON HAWTHORN.

even in a picturesque point of view, is not only an interesting object by itself, but produces a most interesting combination, or contrast, as things may be, when grouped with other trees.

"We have seen it hanging over rocks, with deep shadows under its foliage, or shooting from their sides, in the most fantastic forms, as if to gaze at its image in the deep pool below. We have seen it contrasting its tender germ, and its delicate leaves, with the brighter and deeper masses of the Holly and the Alder. We have seen it growing under the shelter, though not the shade, of some stately Oak, embodying the idea of beauty protected by strength. Our eyes have often caught the motion of the busy mill-wheel, over which its blossoms were clustering. We have seen it growing grandly on the green of the village school, the great object of general attraction to the young urchins, who played in idle groups about its roots, and perhaps the only thing remaining to be recognized when the school-boy returns as the man. We have seen its aged boughs overshadowing one half of some peaceful woodland cottage, its foliage half concealing the window, whence the sounds of happy content and cheerful mirth came forth.

"We know that lively season,

> 'When the milkmaid singeth blythe,
> And the mower whets his scythe,
> And every shepherd tells his tale
> Under the Hawthorn in the dale.'

And with these, and a thousand such associations as these, we cannot but feel emotions of no ordinary nature when we behold this beautiful tree."

In another place Gilpin speaks more favourably of the tree in question: "Though as a single bush it is sometimes offensive, yet, entangled with an Oak, or mixing with other trees, it may be beautiful." Price, in his admirable *Essay on the Picturesque*, expresses the same opinion: "Should it happen, for example, that in parts of the rising ground of a highly dressed lawn, groups of *Thorns* and Hollies were mixed with the Oaks and Beeches, is there any one, with the least taste for natural beauties, who would totally extirpate them, and clear round all the larger trees? Is there any one who would not delight in such a mixture?"

If the artist, however, still refuses to admit it into his painted landscape, we must be content to admire it as it stands in the landscape of Nature—to enjoy its rich perfume—to contrast its bright green leaves with its ivory flowers—to inspect its minute beauties—to thank it for the hospitable shelter that it affords to the nightingale in summer, and the welcome repast to many a feathered songster in winter, at which season the robin-redbreast selects its topmost twig from which to pour forth his plaintive yet cheering song—and to rest satisfied, that, although it has found no painter to eulogize it, it has never wanted a poet.

"Amongst the many buds proclaiming May,
Decking the fields in holiday array,
Striving who shall surpass in braverie,
Marke the faire blooming of the Hawthorne tree,
Who finely cloathed in a robe of white,
Fills full the wanton eye with May's delight."

Chaucer.

"Gives not the Hawthorn-bush a sweeter shade
To shepherds looking on their silly sheep,
Than doth a rich embroider'd canopy
To kings that fear their subjects' treachery?"

Shakspeare.

"Come, my Corinna, come, and coming, mark
How each field turns a street, each street a park
Made green and trimmed with trees; see how

Devotion gives each house a bough
Or branch: each porch, each door, ere this,
An ark, a tabernacle is;
Made up of Whitethorn neatly interwove."

Herrick.

"From the moist meadow to the wither'd hill,
Led by the breeze, the vivid verdure runs,
And swells, and deepens, to the cherish'd eye.
The Hawthorn whitens; and the juicy groves
Put forth their buds, unfolding by degrees,
Till the whole leafy forest stands display'd
In full luxuriance to the sighing gales."

Thomson.

"The Hawthorn-bush, with seats beneath the shade
For talking age and whispering lovers made."

Goldsmith.

"From the Whitethorn the May-flower shed
Its dewy fragrance round our head."

Scott.

"The Gorse is yellow on the heath,
The banks with Speedwell flowers are gay;
The Oak is budding, and beneath,
The Hawthorn soon shall wear the wreath,
The silver wreath of May."

Charlotte Smith.

"The milk-white Thorn that scents the evening gale."

Burns.

"Yon reverend Hawthorns, harden'd to the rod
Of winter storms, yet budding cheerfully."

Wordsworth.

The Hawthorn, according to some etymolgists, is so called from its fruit, or *haw:* or, if Booth be correct, the tree gives the name to the fruit; the first syllable of the word being a corruption of *hage*, or *hæg*, and the word itself signifying a *hedge-thorn.**

* Scott, in his *Discovery of Witchcraft*, calls it "Hay-thorn." Cratægus is from the Greek *κϱατος strength:* Oxyacantha signifies *sharp-thorn;* Pyracantha, *fiery-thorn.*

Cratægus and Oxyacantha, to which may be added Pyracantha, are the names by which the

HAWTHORN BLOSSOM.

Greeks are supposed to have designated the tree. By the Romans it appears to have been called Spina. Its French name, Aube-épine, refers to its flowering early in the spring, or morning, of the year; *aube* signifying "the dawn of day." With us it is known indifferently by the names, May-tree, May-bush, from its season of flowering, and from the important place which it held in the old May games; Quickthorn, Quickset, and simply Quick, from its application to the construction of *quick*, or live hedges, instead of dead branches of trees; and White-thorn, from the profusion of its white flowers. By some botanists it is placed in the same genus with "Mespilus," the *Medlar*, with which it has many botanical characters in common.

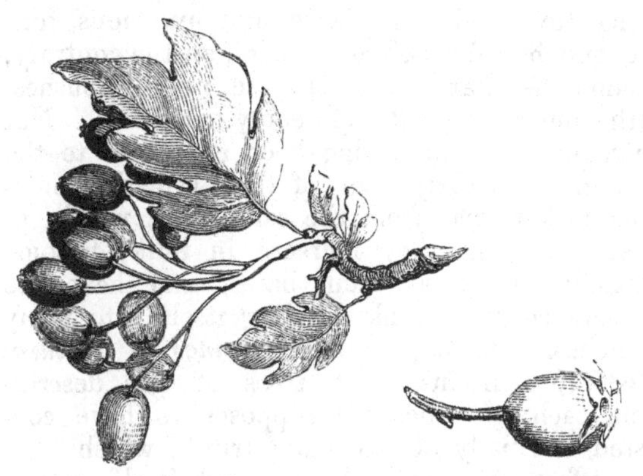

FRUIT OF HAWTHORN. NAT. SIZE.

It is found in most parts of Europe, from the Mediterranean to as far north as $60\frac{1}{2}^{\circ}$, in Sweden; in the north of Africa, and in western Asia. It was introduced many years since into Australia, where it grows as luxuriantly as in its native country, and where it must have no little efficacy in keeping alive the memory of the shady lanes and village greens of Old England.

It would be superfluous for me to give a detailed description of a tree with which every one is so familiar as the Hawthorn. I will therefore simply make a few remarks on its mode of growth and other peculiarities, which I will leave to my readers to verify at their leisure.

In size, mode of growth, foliage, colour, and even odour of its flowers, the Hawthorn is perhaps more liable to variation than any other tree. Some exhibit a strong, free, and upright growth, being furnished with large and luxurious foliage, and but few spines; others, on the contrary, assume the character of stunted, prickly, bushes, with numerous small and deeply cut leaves. Not unfrequently, from having been cut down to the ground in an early stage of their growth, numerous suckers rise from the same root, which, in after years, as they increase in bulk, become partially united at their bases, and have the appearance of a trunk dividing itself into many branches. Jesse, in his *Gleanings of Natural History*, mentions some trees of this description, each of which he supposes to have consisted originally of one main trunk, which from the effects of age had separated itself into a number of smaller stems. "While on the subject of trees," he says, "I will notice the present state of the old Thorns in Bushy Park, from which it probably takes its name. These trees are generally supposed to have been in existence at the time of Oliver Cromwell, the park being then used as a hare park. As they increase in age, they have the property of separating themselves into different stems, some having four or five, or even six, which, as they separate, become regularly barked round, forming, to appearance, so many distinct trees closely planted together, except that they all meet at the butt of the tree. Some of the trees are now undergoing this process of separation, having already thrown out one stem, while in other parts they are deeply indented with seams down the whole stem. These, gradually deepening from opposite sides towards the centre, will at last split the tree into a number of separate stems, which are barked round. In other trees the seam is hardly visible, though none of them are without it. This peculiarity seems confined to the Thorn, and as I have not observed it in those which appear to have been more recently planted, it is probably the effect of great age, though the trees are still flourishing, and I know of few sights more beautiful than the fine old Thorns in Bushy Park in full blossom."

Now, I am inclined to think that, although this description of the Thorns is, no doubt, perfectly correct, the supposition on which Mr. Jesse accounts for the separate stems is not equally so. The easiest and simplest proof would be to cut down one of the stems, and to observe whether the wood is *arranged in concentric circles*, having the pith for a common centre, or, whether the layers of wood are *broken into irregular segments of circles*, without any common centre. If each stem be found to have a distinct central column of pith, with the wood deposited in layers around it, (which, I apprehend, will turn out to be the case,) it must have been a separate stem from the beginning of its existence; for there is never more than one column of pith in the same trunk, and that one was fully formed during the first year's growth, and has never since increased in size. In the latter case only, therefore, can his theory be well founded. Certainly, in some very old trees at Newham and Penrose, near Helston, Cornwall, the appearance of which closely resembles those which he describes, the trees originally consisted of a number of stems which have grown together.* The true explanation I believe to be this. The Hawthorn, when cut down close to the ground, invariably sends up several strong shoots from the buds nearest to the root, which, from their having no room to expand, have a natural tendency as they increase in size to grow together. When once thus united, the lower portion of the consolidated trunk would present the "seamed" appearance described by Mr. Jesse, and as the tree grew old, the increased weight of branches, foliage, and fruit would have the effect of separating the whole into its constituent parts, or, in fact, restoring it to its former condition. This tendency of the Hawthorn to form numerous separate stems is so well known to hedgers, that it is usual to cut down whole hedges of quickset, as soon as the plants are well rooted, for the sake of forming a thicker fence.

Occasionally, but rarely, the Hawthorn assumes a pendant or "weeping" character. There is a fine tree of this kind in the garden which belonged to the Regent Murray in Scotland, and is said to be very beautiful. Like many other trees, the Hawthorn is occasionally liable to an unhealthy mode of growth, when tufts or clusters of twigs are produced, resembling, if observed at a little distance, a large bird's nest. Mr. Anderson, the curator of the Chelsea Botanic Garden, had the curiosity to graft young Thorns with

* Grigor, in his *Eastern Arboretum*, describes a similar tree at Earsham, Norfolk. "the trunk of which was a series of stems massed and matted together, measuring, at five feet from the ground, nine feet in circumference."

some of these twigs, and found, in the course of two or three years, that they produced beautiful weeping branches.*

It has already been said that the varieties of the Hawthorn are very numerous, and no less strongly marked. Difference of soil and situation produces yet more remarkable contrasts. A bushy tree in the rich lowlands, it becomes gnarled, ragged, and fantastic in form, as it creeps up the mountains, and finally dwindles into a mere stunted and knotty shrub.

> "There is a Thorn—it looks so old—
> In truth, you'd find it hard to say,
> How it could ever have been young,
> It looks so old and grey.
> Not higher than a two years' child,
> It stands erect, this aged Thorn;
> No leaves it has, no prickly points;
> It is a mass of knotted joints,
> A wretched thing forlorn.
> It stands erect, and like a stone
> With lichens it is overgrown.
>
> Like rock or stone, it is o'ergrown
> With lichens to the very top,
> And hung with heavy tufts of moss,
> A melancholy crop:
> Up from the earth these mosses creep,
> And this poor Thorn they clasp it round,
> So close, you'd say that they are bent
> With plain and manifest intent,
> To drag it to the ground;
> And all have joined in one endeavour,
> To bury this poor Thorn for ever."
>
> *Wordsworth.*

It is difficult to imagine that this "Thorn so old and grey," had it stood on the village-green beneath, might have found a place in the same

* Similar results followed from budding, or grafting, from the tufts produced by the Elm (*Ulmus campestris*).

P

poet's song as the tree under whose shade

> "shepherds sate of yore and wove
> May-garlands."

The spines, or thorns, which form a characteristic feature of this tree, are to be distinguished from *prickles*, such as those which invest the stems of the rose or bramble. The latter are attached only to the surface of the stem, and even to that sometimes not very firmly. Thorns, however, are to be considered as imperfect branches, being furnished with proper bark, wood, and pith of their own. They enlarge in the second year of their growth, and for the most part produce buds and leaves, and eventually flowers and fruit; whereas prickles never increase in size after the first year, and are not converted into branches.

Not even is the colour of the blossom which gives the name "White-thorn" free from variations. Indeed, most commonly, it assumes a pink hue in fading; but in gardens and shrubberies, varieties are frequent in which the flower is of a permanent and decided pink or crimson. The perfume of the blossom is generally exceedingly fragrant; but occasionally this fragrance is almost overpowered by a strong fishy smell, which is most perceptible when the branch is held close to the nose, or carried into a close room. The *haw*, too, varies greatly in size, shape, and colour, being sometimes oblong, sometimes nearly globular; sometimes downy, at other times smooth and polished. Varieties have been observed in which it exchanges its usual crimson hue for black, orange, golden yellow, or white. In the West of England, and probably most other parts of the country, each haw contains a single nut: but in the neighbourhood of Barnet and Hadley, in Hertfordshire, I have observed that they more frequently contain two.

The pink and double varieties of Hawthorn are multiplied by grafting and budding, but the common sort is generally raised from seed. The haws are gathered in winter and laid in a heap, mixed with a sufficient quantity of soil to cover them and separate them from each other, and exposed to the influence of the weather, until the spring of the second, or even the third ensuing year. Unless this plan is adopted, the young plants do not appear till the year after they are sown, and consequently occasion the loss of the ground for that time. Various experiments have been tried with the seed, in the hope of finding some method of securing their growth in the year following that of their being gathered, but none have succeeded. The extreme hardness and durability of the shell is the probable cause of this sluggishness of growth. Could any plan be devised for breaking the shell without injuring the kernel, it is not unlikely that the desired object would be effected.

I have already spoken of the claims of the Hawthorn to picturesque beauty. Whether they are allowed or not, there can be no doubt that not only the several varieties of the British tree, but many foreign species, are eminently ornamental to the lawn and shrubbery.

In husbandry, the principal use of the Quickthorn is for making hedges, for which purpose very many thousands are annually raised in Britain, an employment which forms an important branch of the business of nursery-men.

This raising of Thorns for profit is a comparatively modern occupation, Evelyn being the first to tell us of a gentleman who had "considerably improved his revenue by sowing Haws only, and raising nurseries of Quicksets, which he sells by the hundred far and near." In the first year of their growth, the seedlings attain the height of from six to twelve inches, and during the two or three following years increase at the annual rate of from one foot to three feet; afterwards they grow more slowly till they have attained the height of from twelve to fifteen feet, when the shoots are produced principally in a lateral direction. This peculiarity, added to the rigidity of its thorns, makes it so valuable for the purpose above-mentioned, the denseness of its side branches being greatly promoted by frequent prunings of the upward shoots. In order to ensure an uniformly dense hedge, the best plan is to plant three or four-years-old trees in two rows, about a foot or a foot and a half apart, and in the following season to cut them down within an inch or two of the ground. If kept clear of weeds, they will make numerous strong shoots during the succeeding year, and soon form an impenetrable barrier. Hedges of this tree will stand the sea-breeze better than most others; but still are far from being uninjured by their rude visitor, for

> "Where from sea-blasts the Hawthorns lean,
> And hoary dews are slow to melt,"

the side most exposed to the weather may frequently be observed rounded off as neatly as if by the gardener's shears. This effect is produced by the particles of salt with which the sea-breeze is charged being arrested by the twigs and killing the young buds: but the opposite side flourishes with tolerable luxuriance.*

Thorns are occasionally liable to attacks from a fungus (*Œcidium laceratum*) which produces singular brown swellings on the young shoots and leaves. Their most usual shape is oval, and in size they vary from that of a bean to that of a walnut. On the outside they are smooth, but internally contain a large quantity of brown powder, which rises in a cloud when the hedge is shaken. Young plants are also liable in damp seasons to a destructive mildew, as a remedy against which strong stimulating manures are recommended, and the application of soot.

The stock of the Thorn is employed not only for grafting varieties of its own species, but also, and with great advantage, for several of the garden fruits.

> "Man does the savage Hawthorn teach
> To bear the medlar and the pear;
> He bids the rustic Plum to rear
> A noble trunk and be a Peach."
>
> *Cowley.*

* Some few years ago, a gardener, accustomed only to the midland counties, was engaged by a gentleman, whose estate lies on the northern sea-coast of Devonshire, to superintend his garden and plantations. On his arrival he was sent by his employer to walk through his domain, that he might gain some notion of what would be required of him. His inspection being completed, he was asked what he thought of his new employment: "I like the place well," he replied, "and doubt not that I should be able to give satisfaction, except on one point. How my predecessor contrived to keep the Thorn-hedges so neatly clipped with only four hands to help him, I cannot tell, nor can I undertake to do as well: I must therefore decline the situation." He was not a little surprised on being told that the north-west wind was his "predecessor," a coadjutor whose services he probably afterwards found verging on the officious.

The leaves, like those of the Beech and some other trees, are invested with a short downy covering while young, which afterwards almost entirely disappears, leaving a bright and glossy surface. They are said to be used not unfrequently for the purpose of adulterating tea; and indeed, not many years since, a patent was taken out for preparing them as a substitute for the more costly leaf; cattle will browze on them, not forgetting to pay due regard to the sharp spines with which the younger branches are plentifully armed.

Most of my readers will, I doubt not, be able to recall a period of their lives when a twig of Hawthorn, just coming into bud, straight and smooth, and furnished with a regular array of unmutilated spines, contributed not a little to the innocent enjoyment of childhood. With a daisy, chosen from among a thousand for its petals deeply tinged with crimson—many times thrown away to give place to another yet more beautiful—and a half-open buttercup, stuck on alternate thorns—who so happy as we were then! When have we been so happy since? Never, perhaps, except when we have stolen away from the world and thoughts of the world, and burying ourselves in the depths of a forest, have discovered, in the solemn and mute aspirations of created nature, intimations of the spring of an Eternity. Never man spake as He did who bade us look to children for our first and most perfect lessons in Christian philosophy: it cannot, therefore, be unattended with profit to ourselves, if, in mature age, we can now and then catch but a transient impression of the feelings which thronged upon our life in childhood—

> "The time when meadow, grove, and stream,
> The earth, and every common sight,
> To us did seem
> Apparelled in celestial light,—
> The glory and the freshness of a dream."

I may here observe that the larger spines of the Hawthorn may be applied, as in some places they are, to a use which ought to be generally known. Poor children in the neighbourhood of towns might be taught to prepare them for the purpose, and almost to gain themselves a livelihood. I shall probably raise a smile when I add that a small amount of labour will convert them into excellent *skewers* for purposes to which pins are more frequently applied. Thorns should be selected about two or three inches in length, as free from knots as possible, and boiled in water for a few minutes. The rind may then be easily removed from every part of them but the point, and they require no further preparation before they are fit for use. Poor children, having first received permission (which scarcely any one would refuse) from the owners of Thorn hedges, might thus be put in the way of benefiting themselves without doing injury to any one. Threepence or fourpence a hundred would be a remunerating price to them; and one which any one would prefer to pay rather than run the risk of inadvertently swallowing a pin. The thorns may readily be slipped from the branch without the aid of a knife; but whoever tries the experiment should provide the child with a common knife to assist in removing the rind, and a sheet of fine sand paper to remove inequalities in the surface.*

With the exception that a strong fermented liquor may be made from haws, neither the blos-

THE HAW-FINCH.

som nor the fruit have been applied to any important use by man: but the flowers as well as the leaves afford sustenance to a variety of insects; and the haws, which are followed, as to the time of ripening, by the berries of the Ivy, and those again by the berries of the Misletoe, produce an abundant supply of food to the feathered tribe during

* I am indebted for this idea to a letter in the Gardener's Magazine, vii. 234, by the Rev. T. W. Bree.

the severest and most protracted of our winters. One bird of passage, the Haw-finch, only remains with us during the months when its favourite food, from which it derives its name, is to be procured, coming in Autumn and departing in the month of April. It has been suggested that pigs might be fed with haws during seasons of scarcity.

The Hawthorn attains a great age, and, when large enough to rank among timber-trees, is of considerable value. According to Evelyn, "The root of an old Thorn is excellent both for boxes and combs, and is curiously and naturally wrought: I have read that they make ribs to some small boats or vessels with the White Thorn; and it is certain, that if they were planted single, and in standards, where they might be safe, they would rise into large bodied trees in time, and be of excellent use for the turner, not inferior to Box." Loudon says "its wood is very hard and difficult to work: its colour is white, but with a yellowish tinge; its grain is fine, and it takes a beautiful polish; but it is not much used in the arts, because it is seldom found of sufficient size, and is besides apt to warp. It weighs, when green, sixty-eight pounds twelve ounces per cubic foot; and when dry, fifty-seven pounds five ounces. It contracts, by drying, one-eighth of its bulk. It is employed for the handles of hammers, the teeth of mill-wheels, for flails and mallets, and, when heated at the fire, for canes and walking-sticks. The branches are used in the country for heating ovens; a purpose for which they are very proper, as they give out much heat, and, like the Ash and Furze, possess the property of burning as readily when green as in their dry state."

It has also been stated that it might be substituted for Box-wood as a material for wood-engraving, in case of any deficiency in the supply of the preferable, but more costly wood. It is often spoiled through inattention after cutting; if it be allowed to lie in entire logs or trunks, it soon heats and becomes quite brittle and worthless; it ought, therefore, to be cut up immediately into planks, and laid to dry.

Remarkable trees of this species are—one mentioned by Jesse, in Dulham Park, Suffolk, which is well worthy of notice, from its great size, antiquity, and the curious manner in which it grows; one at Kinkarochie, in the parish of Scone, in Perthshire, which in 1795 measured nine feet in circumference; another at Duddingstone, in the county of Edinburgh, which measures nine feet at three feet above the ground, and a little way above the roots, twelve feet round. These two last are mentioned by Sir T. D. Lauder.

But the most remarkable tree, next to the Glastonbury Thorn already noticed, is the Hethel Thorn, of which the following account is given by Grigor in his *Eastern Arboretum*.

"It stands in a field adjoining the church, the property of Hudson Gurney, Esq., and though it still greets the May-morn with its profuse and odoriferous blossoms, and bears a plentiful crop of fruit like the others, it is invested with a character differing materially from that of the species in general, arising from its extreme old age. In looking upon it, one would suppose it had been here for thousands of years; and indeed, if the common tradition of the place is to be relied upon, it must be acknowledged to be, in a high

THE HETHEL THORN.

degree, patriarchal. Participating in the general interest felt with reference to this object, we deemed it advisable to apply for information regarding it to the proprietor, to whom we are indebted for the following remarks, which we take the liberty of here introducing:—As to the Hethel Thorn, 'I wish,' he says, 'my story were a clearer one, and should be very glad if authorities and traditions could be better collected. I have heard that the first Sir Thomas Beevor said, that he was in possession of a deed bearing date early in the thirteenth century, in which, referring to it as a boundary tree, it is mentioned as the Old Thorn. But I have innumerable deeds from the court-rolls of the manor of Hethel, but none of them earlier than thc time of Edward III., and amongst them I can find no such mention. If, therefore, Sir Thomas had such deed, he must have taken it out, and kept it as a curiosity. I have also heard that in one of the chronicles the Thorn was mentioned as the mark for meeting, in an insurrection of the peasants in the reign of King John; but I have never been able to get a reference to what chronicle. The first Sir Thomas Beevor put a rail round it, and took great care of it. After the present Sir Thomas Beevor left Hethel, it was much neglected: and pieces of it, I am told, were pulled down by the cottagers. I replaced the railing, and had some of the branches, which had been supported by crooks by Sir Thomas, again so propped. Not only the bark of the hollow tree is as hard and as heavy as iron, but every branch, most curiously inter-involved, is a hollow tube, into which you may put your arm, all the interior wood being gone. It still puts forth blossoms and haws everywhere yearly, but I think during later years it has been sensibly going off, and is not the tree it was even five years ago.'

"Subsequently to our receiving this information, we spent part of a day amongst the villagers of Hethel, for the purpose of collecting any traditions that might be retained regarding this tree; but we had the satisfaction only to know that what is embodied in the foregoing statement is all that we could learn from them regarding it. The older inhabitants with whom we conversed, some of them nearly ninety years old, ascribe to it an extraordinary age, and consider it, very justly, we think, to have been the glory of all the Thorns in the neighbourhood for many centuries.

"This, however, is saying but little; for its very appearance justifies us in allotting to it an age of more than five centuries, whilst its size as a Thorn is remarkable. Our measurement of it stands as follows:—at one foot from the base of the trunk, 12 feet 1 inch in circumference, and at 5 feet high, 14 feet 3 inches; whilst the circumference of the space over which the branches spread is 31 yards. Its trunk is reduced to a mere shell, and though somewhat divided, it has none of that shattered appearance which we sometimes observe in the Oak. The ramification of the top had assumed a style which we can neither trace in the Oak, nor in any trees of its own species, the branches forming a thick grotesque mass most curiously interwoven. It is covered all over with lichen, and crowned with misletoe, adding still more to the effect which age confers upon such objects. Whether by accident or otherwise, we know not, but it appears to have lost several boughs of late years, so that it is not nearly of such a large size, as a spreading tree, as it was even ten years ago. The boys of the village, too, are in the habit of going a-Maying to the 'Old Thorn,' and robbing it of large bunches of spray every season, a practice which, as to this particular tree, should be at once discontinued. It appears that Mr. Marsham, in a communication to the Bath Society, nearly a century since, has noticed this Thorn as a remarkable tree, and stated its measurement at 4 feet from the ground to be 9 feet 1 inch; remarking also, that one of its arms extended above seven yards. Whatever, then, may be the correct age of this tree, it is unquestionably the most interesting specimen in the east of England, and fairly entitled, we think, to rank amongst the most celebrated in our country."

Through the kindness of the Rev. J. H. Steward, I am enabled to give the following additional particulars respecting the present state of the tree.

"The present proprietor of the field in which the Thorn stands, (who purchased the property of the present Sir Thomas Beevor,) is an antiquary, and I have no doubt has ascertained all that can be ascertained of the traditionary history of the tree. In Mr. Grigor's book there is an etching, which gives very nearly the present appearance of the Thorn. I have to-day had the tree measured, and at 4 feet from the ground, its girth is 12 feet 5 inches; this excess over the girth given in Loudon appears to have been occasioned by the splitting and giving way of the trunk, and not by growth. Loudon, quoting Marsham, says, 'One arm of it extending above 7 yards:'—there is now a branch which extends 6 yards. At 5 feet from the ground the girth is 16 feet 3 inches, an excess beyond Grigor's measurement to be accounted for as above stated. At 1 foot from the ground, it is 12 feet 2 inches. —The present circumference of the branches is 31 yards; the longest diameter of them, 10 yards and 2 feet, and the height about 18 feet."

Grigor also mentions another interesting Thorn, which stands in the churchyard of St. Michael's at Thorn, Norwich, and which bears the marks of great antiquity. Though situated in a thickly populous district of the city, blackened with dust, it is found every year at its appointed time mantled with sweet *May*. It has undergone little alteration in appearance during the last half-century; and, as some kind hand has placed a hoop of iron round its shattered stem, it bids fair to withstand the inroads of time for many years to come.

In the mountainous parts of Derbyshire, called Woodlands, stands a Hawthorn which affords a curious instance of the provision made by nature for the support of a decayed tree. It once had a large and stout trunk, but at the time when it was observed was quite hollow, both wood and bark being decayed and lifeless; but the foliage was supported in full vigour by a root about three inches in diameter, which had descended through the hollow and performed the functions of a stem.

WITCH THORN.

The following account of a remarkable Thorn in the north of Ireland, together with the annexed sketch, was supplied by Alexander Johns, Esq.

"On the estate of M. Dalway, Esq. at Bella Hill, county of Antrim, stands the 'Witch Thorn,' but I have not been able to trace why so called, nor to collect any legends respecting it. The schoolmaster of the Witch-Thorn National School (the tree has given its name to the place) referred me to an old man named James Poag, residing about a quarter of a mile from the spot. I found him at home, but gained little information; he is eighty-seven years of age, a tailor by trade, and was busy at his work, three lads plying the needle with him: he said his sight was not so good as it had been, and his hearing rather dull! He invited me to take bread and butter and milk, all his house afforded, and told me he remembers the tree for seventy years, and that from his earliest recollection the trunk has always been as large as it is now. Within these few years some branches have been cut off, (a very rare occurrence indeed with an aged Thorn,) which being reported to the agent of Mr. Dalway, that gentleman went to the spot, and has taken steps to prevent a repetition of the act. The large trunk is 4 feet 2 inches in circumference, and the other 3 feet 6 inches; the Thorn is about 20 feet high. It stands on high ground, and the father of the present proprietor told my informant that he had seen the Witch Thorn from the Scotch coast.

"These old Thorns are numerous in Ireland, and are greatly venerated. They are sometimes found in fields in crop, but in ploughing, care is taken not to approach the roots, lest injury should be done. I have seen old Thorns near wells almost covered with bits of garments, particularly in the south of Ireland."

Among the numerous insects which either wholly or in part derive their sustenance from the Hawthorn, I will proceed to describe the most worthy of note.

The Black-veined white, or Hawthorn Butterfly, (*Pieris Cratægi*,) derives its name from one of the favourite trees to which it resorts to lay its eggs,

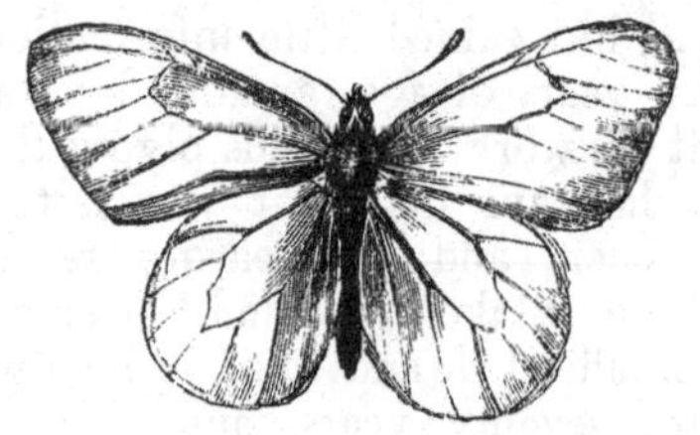

HAWTHORN BUTTERFLY.

and on which its caterpillar feeds. The perfect insect resembles very closely the Cabbage Butterfly (*Pontia Brassicæ*), so destructive in kitchen-gardens.

It differs principally from the latter in having the ribs or veins of the wings black, and in being destitute of the two black spots on the under-side of the fore wings, which characterise the Cabbage Butterfly. In the month of June or July, the Hawthorn Butterfly lays from a hundred to a hundred and fifty yellow, cylindrical, ribbed eggs, on a leaf of the Hawthorn, apple, or some other tree of the same tribe, selecting generally the tallest trees. The eggs lie exposed upon the leaf without being covered with any sort of wool, and after the lapse of about a fortnight become of a silvery colour, and are more deeply ribbed. The caterpillars can now be discovered shining through some of them, and in the course of two or three days more they all burst their shells and commence their predatory life, the empty cases melting like wax in the heat of the sun. The newly hatched caterpillars are of a dirty yellow colour, and covered with hair; the head is black, and there is a black ring round the neck, and a brownish red stripe on both sides of the body. On the approach of rain they draw the leaf together over them by means of a web, and thus secure for themselves a sufficient shelter, however unfavourable the weather may be. This living penthouse serves them not only for shelter, but for food; in the course of a few days, therefore, having devoured the whole of the leaf except the veins, they find it necessary to look out for a new lodging, and add the next leaf to their old abode, connecting it by a web. But as it sometimes happens that summer-storms have the same effect as autumnal frosts in stripping the trees of their foliage, in order to prevent any such catastrophe befalling their dwelling, they secure the leaf under which they have taken shelter by fastening it to the shoots with threads. In rainy weather, or when the sun is very hot, they remain quiet at home; but as their appetite increases with their size, and the walls and roof of their dwelling-house still constitute their only subsistence, in two days' time another change of residence is necessary, and is effected in the same way. But easy as they find it to provide themselves with food and lodging, they are not exempt from misfortune. Birds and insects destroy them, and many perish from unfavourable weather. The family thus rapidly dwindles away to twenty or thirty, so that a single shoot generally furnishes them with sufficient food for the summer. Early in autumn they seem to be endowed with a greater amount of forethought: nipping frosts are at hand, which their delicate structure is not prepared to encounter; cold wintry blasts are about to set in, which, unless they make a strenuous effort to prevent it, will whirl away both themselves and their habitations. In September, therefore, they cease eating, and set about preparing a winter's habitation. They bend one leaf over another, bring the edges close together, and unite them with webs, covering the chamber thus formed within with a fine web, so as only to leave themselves a small space to enter at. They also unite the leaf-stalk which they have prepared for their nest with the shoot, so that neither wind nor rain can detach it, and, this operation completed, they all return to the nest, and secure it on all sides from wet and cold. So far, all their proceedings have been conducive to the public good; now, however, private comfort is to be attended to: each caterpillar selects a place in the chamber, in which to pass the dreary months of winter, and envelopes himself in a silken web which defies the action of the severest weather.

New life and new instinct return with the first warm days of spring. The sunshine entices a few of the caterpillars out of their nest; but, as if doubtful whether they can as yet duly supply themselves with food, they retire to their dwelling without tasting a bud. In the beginning of April, the temperature being decidedly milder, they sally forth in a body, and, finding the blossom-buds swelling and beginning to shoot, make up for their long fast by devouring them. The leaf-buds now begin to expand. These they attack in their turn, discard their decayed winter's dwelling, and construct a larger and more commodious mansion, to which they retire every evening, or during inclement weather. After their second change of skin, which takes place about this time, they grow very rapidly, and alter in appearance, having now a black line running along the centre of their backs, with a row of yellow spots on each side. Their backs are covered with yellow and white hairs, and from the central line grey stripes diverge to the under side of the body. Towards the end of April they moult for the third time, and dissolve the republican state of society. Each caterpillar now attaches itself by threads to a branch of the tree, and in the course of a few days is converted into a chrysalis, or pupa, which is of a whitish yellow colour, beset with black dots and stripes. It remains in this state till the end of May or beginning of June, when the perfect insect appears, and soon lays the foundation of another colony.

Providentially, this destructive insect has a number of enemies, or our Hawthorn-hedges and orchards would present but a pitiable appearance. Field-bugs watch the caterpillars when they leave the nest, pierce them with their beaks, and suck the juices. Ichneumon-flies lay their eggs in the bodies of the full-grown caterpillars, which afterwards serve as food for the ravenous larvæ of the parasitic insect. Small birds, especially the much calumniated, but really valuable, tom-tit, eagerly devour them soon after they are hatched, as well as in the following spring when they are dispersed upon the shoots, and even break open and rifle the habitations in which they had so carefully ensconced themselves for the winter. Owing to these united causes, the number of those which survive till the spring is very limited; so that it is only when their natural enemies, the birds and above-mentioned insects, have been destroyed, either by natural causes or by human agency, that their ravages are seriously to be apprehended.

Another insect which frequents the Hawthorn and several of our forest and fruit trees, is the "Yellow-tailed Moth" (*Bombyx*, or *Porthesia*, *chrysorhœa*). The caterpillar of this insect closely resembles that of the Hawthorn-butterfly, differing principally in having two reddish-yellow tubercles near the extremity of the body, and four near the head. Its habits, too, are very similar: but its winter quarters are somewhat more complicated, consisting of several leaves, and divided into chambers. The caterpillars of this species are preyed on by small birds to a less extent than those of the last-mentioned, owing, it is conjectured, to the thick hairs on their backs; but their excessive increase is checked by a very small fly, scarcely visible to the human eye, which lays its eggs singly in the eggs of the Yellow-tailed Moth, so that instead of a caterpillar, the larva of a fly, is produced, which makes use of the egg for its food, and the shell for its dwelling. Field-bugs and Ichneumon-flies are also appointed instruments for checking their ravages; and

YELLOW-TAILED MOTH.

the Moth itself is frequently washed from the tree while laying its eggs, and perishes on the ground.

Another species of the same genus, the "Brown-tailed Moth" (*Bombyx*, or *Porthesia*, *auriflua*), has occasionally been exceedingly destructive.

In the summer of 1782 especially, this insect created great alarm over the country, from its colonies established on the Hawthorn and some other trees, which were so exceedingly numerous, we are informed by Mr. W. Curtis (who published a treatise on it), that in many of the parishes near London subscriptions were opened, and the poor people employed to cut off the webs *at one shilling per bushel,* which were burnt under the inspection of the churchwardens, overseers, or beadles of the parish; and some idea may be formed of their numbers from the fact, that at the outset of this business *eighty bushels were collected in one day in the parish of Clapham alone.* The mischief, however, was not confined

BROWN-TAILED MOTH.

to the actual destruction of the vegetation, but the most absurd alarms were raised from the sudden appearance of these creatures, which by some were regarded as the forerunners of the Plague; by others, as the actual *cause* of it; and by some it was supposed that the destruction of every kind of vegetable would follow. Prayers were offered up in some of the churches to deliver the country from the apprehended approaching calamity.

We learn from Holy Writ, as well as from the testimony of travellers, both ancient and modern, that an instrument, apparently so contemptible as an insect, is occasionally employed by the Almighty as a national scourge — that creatures, whose agency, when they are taken singly, can scarcely be said to be productive of any appreciable effect, are sometimes commissioned to spread famine and desolation to a degree not to be surpassed by the worst horrors of war. We know, too, that the produce of the earth may be checked, and the hopes of man disappointed by the instrumentality of a yet meaner agent. Witness the invisible and unknown cause, which, in the present year (1846), defying all the theories of our wisest philosophers, is secretly acting on our potato-fields, and depriving a large portion of our rustic population of a staple article of food. We cannot think on these things without reflecting on the unsearchableness of the ways of God, or without deriving to ourselves a deeply practical lesson in humility. We may, if we will, gain yet further instruction from the history of the Hawthorn Butterfly. The transition of insects generally from the *chrysalis* to the *perfect* state has been compared, ages ago, to the resurrection, when the redeemed shall rise from their earthly tombs with glorified bodies; and it would be rash to pronounce the comparison fanciful, for Solomon himself has referred us to the insect world for wisdom,* as a wiser than he has directed our attention to the vegetable. When, therefore, in the bright joyous days of June, we see so common an object as a white butterfly flit by us, we may reflect with advantage on the period of difficulty and danger, of watching and painstaking, that it completed before it passed into its grave—the chrysalis, soon to emerge endued with a new body, new appetites, and new powers. Is it presumption to say that God has so ordered its ways that it might furnish us with an example not merely of worldly, but of heavenly prudence? I believe not: the devotional and faithful observer of the objects of sense sees in *all* the works of Nature the works of God, and must not be condemned if he can derive from the few, which he is permitted to a limited degree to comprehend, inducements to prepare himself for his own approaching change, with the patience, industry, and watchfulness of an insect, which, obedient to the dictates of instinct, has completed the grovelling stage of its existence, as God intended that it should, and has reached its destination. And, after all, the poor worm had only instinct for its guide, and its perfect state was but a butterfly's life. We have reason and Revelation, and the Holy Spirit, for *our* guides, and our perfect state is for eternity; yet how many of us would reverse the order of God's Providence, and are bent on leading the butterfly's life first!

* "Go to the ant, thou sluggard, consider her ways and be wise." Prov. vi. 6.

SLOE FLOWER.

THE BLACKTHORN.

PRUNUS SPINOSA.

Natural Order—ROSACEÆ.

Class—ICOSANDRIA. *Order*—MONOGYNIA.

THE subject of the last chapter has high claims to be ranked among the most interesting of British trees, being not only a beautiful ornament to the landscape at all seasons, but possessing a legendary character which secures for it more than the passing attention of the antiquary. The subject of the present memoir, however, though its name might lead us to hope that it had more points of resemblance to the Hawthorn, possesses but little interest for botanist, forester, painter, or antiquary. In its natural state it is a rigid, wiry bush, remarkable for no beauty of flower or foliage, and not making up for its outward deficiencies by any inherent virtues residing in fruit, stem, or root.

Its very flowers, which are numerous and appear early in Spring, can barely be called ornamental. Expanding, as they do, before any other tree has ventured to show signs of returning life, we are inclined to look on them in the light of daring adventurers, rather than harbingers of the time which "purples all the ground with vernal flowers." Their white ragged petals contrast strangely with the sombre hues of the bare boughs around them — they look cold and cheerless, and carry the mind back to the frosts and snow of the winter which has just passed, instead of forward to the bright days of spring, which are coming. A single primrose, a leaf-bud of Hawthorn or Elm—either of these is a prophet in whom we place unbounded confidence; they are emblems of soft west winds and sunny showers: but the Black-thorn bespeaks our attention to the possible return of *black* east winds, frosty nights, and nipping blights.*

Nor does the Sloe-tree find a champion in the husbandman. It is by no means particular in its choice of soil and situation, but thrives everywhere. Its long creeping roots extend so rapidly, that in the course of a few years a single plant would, if left unmolested, cover an acre of ground.† Thus left to itself, it has no disposition to assume the character of a tree, but forms a low thicket, to the exclusion of every more valuable plant, and, if growing in the neighbourhood of sheep-walks, most unceremoniously levies contributions from every woolly visitor who comes within reach of its knotted and thorny branches. If, by being

* "This tree usually blossoms while cold, north-east winds blow; so that the harsh, rugged weather obtaining at this season is called by country people, 'Black-thorn winter.'"—*White's Selborne.*

† "The name of Mère-du-Bois (*Mother of the Wood*,) is applied to the Sloe-thorn in France, in the neighbourhood of Montargis, because it has been remarked there, that when it was established on the margins of woods, its underground shoots, and the suckers which sprung up from them, had a constant tendency to extend the wood over the adjoining fields; and that, if the proprietors of lands adjoining forests where the Sloe-thorn formed the boundary, did not take the precaution of stopping the progress of its roots, these would, in a short time, spread over their property; and the suckers which arose from them, by affording protection to the seeds of timber trees, (which would be deposited among them by the wind, or by birds,) would ultimately, and at no great distance of time, cause the whole to be covered with forests."—*Loudon.*

deprived of its suckers, it is compelled to throw all its strength upwards, it will sometimes attain the height of thirty feet, and even in natural situations, where it cannot extend itself laterally it rises to fifteen or twenty feet. The name "Blackthorn" appears to have been given to it from the hue of its bark, which being much darker than that of the Hawthorn, probably originated the name of "White-thorn" given to the latter tree.

SLOE.

The *epidermis*, or outer coating of the bark, has, in this species, as in most others of the same genus, a tendency to split horizontally, and to curl back while yet partially attached to the tree.

The leaf is small, of a dark green colour, slightly downy underneath, especially at the junction of the veins, and in its young state. The flowers are white, and conspicuous only from their abundance; as they expand before the leaves, and are consequently unrelieved by any verdure, they are not beautiful. The fruit when ripe is black, and being covered with a delicate bloom, presents, late in the autumn, a more pleasing appearance than the tree can display at any other season.

It is found throughout Europe, with the exception of the extreme north; it occurs also in the north of Africa, and many parts of Asia, and has been introduced into America, where it is frequently found in hedges perfectly naturalised.

The Blackthorn is not nearly so valuable for the construction of live-hedges as the Hawthorn, owing, in the first place, to its rambling habits; and, secondly, to its tendency to send up perpendicular branches, which are bare of thorns towards the base. The wood rarely attains a size which will allow it to be applied to any useful purposes as timber; but the straight stems are extensively used as walking sticks, which are much admired for their bright colour and numerous knots. The thorny dead branches are also recommended as being well adapted for forming a fence round young trees planted in parks, the sharp and rigid thorns effectually preventing the inroads of cattle. The leaves are used to adulterate tea, for which they form a substitute less liable to detection than almost any other British plant, possessing a bitter, aromatic principle, which, inasmuch as it is to be attributed to the presence of prussic acid, must render them very unwholesome. The fruit is intensely austere and astringent, so much so that a single drop of the juice placed on the tongue will produce a roughness on the throat and palate which is perceptible for a long time. When mellowed by frost, however, it becomes red and pulpy, but at no period of its existence claims to be considered a grateful fruit. The juice of it, in its unripe state, is said to enter largely into the composition of spurious port-wine, and it may, it is said, be fermented into a liquor resembling new port.

So impudently and notoriously is this fraud carried on in London, and so boldly is it avowed, that there are books published called "Publican's Guides," &c., in which receipts are given for the manufacture of port-wine from cider, brandy, and sloe-juice, coloured with tincture of red sandars or cudbear.* This villanous compound may be converted into "old port" in a few days by the addition of catechu. The corks may be stained by being soaked in a strong decoction of brazil-wood and a little alum; and even bottles are manufactured which contain a sufficient quantity of lime to be sensibly acted on by the acid, and produce a counterfeit "crust!"

In France the unripe fruit is sometimes pickled and sent to table as a substitute for olives, and in Germany and Russia it is crushed and fer-

* Red-sandars is a preparation of sandal-wood, used as a dye. Cudbear, so called after a Mr. Cuthbert, who first brought it into use, is a lichen, (*Lecanora tartarea*,) found growing in several parts of the Continent, and in Great Britain, on granitic and volcanic rocks, and is also used as a dye. The chemical test called *litmus* is a preparation of this vegetable. Catechu is a substance procured by boiling chips of the heart-wood of *Acacia Catechu*; it is a dark-coloured, powerful, astringent.

mented with water, and a spirit distilled from it. In Dauphiné, the juice of the ripe fruit is used for colouring wine. Letters marked on linen or woollen with this juice will not wash out. The substance sold by druggists under the name of *German Acacia* is prepared from the juice of the unripe fruit.*

The bark, according to Dr. Lindley, is one of the substances which has been reported to resemble "Jesuit's bark"† in its effects. It may be used for tanning leather; a decoction of it with alkali dyes yellow, and it may be employed with advantage as a substitute for galls in the manufacture of ink. There are several varieties, differing principally in the size of the leaf and fruit; but the only one deserving notice is the double flowered, which is cultivated and said to be highly prized in Japan and China for the abundance of its blossom.

* The *true*, or *Egyptian*, *Acacia* is the production of *Acacia Nilotica*, and is used in medicine as a mild astringent.

† Jesuit's bark, Peruvian bark, or (as it is simply called) bark, is the produce of several species of trees growing in Peru, and belonging to the genus *Cinchona*. The order to which these trees belong, *Cinchonaceæ*, is remarkable for containing a large number of plants, the medicinal properties of which render them highly valuable to man. Many of them partake largely of the properties residing in Peruvian bark; Ipecachuana, on many accounts, is a valuable ally of the physician, and several other genera possess similar virtues; Coffee belongs to the same order; while the properties of some are so active that they are said to be a certain antidote against the bite of serpents, and others are so deleterious in their effects that Indians have been poisoned by using the wood to make spits for roasting meat upon, while others again are employed for the destruction of rats and mice. Jesuit's bark received its name from having been introduced into notice by Cardinal de Lugo and the Jesuits about the middle of the seventeenth century. The tree which produces it derived its name "Cinchona" from the lady of the Spanish Viceroy, the Countess del Cinchon, who was cured by the bark. According to some, the Peru-

On the whole, the Blackthorn, in its natural state, possesses few valuable qualities. It certainly does not recommend itself to our favourable consideration on the score of beauty, and being employed to adulterate some substances, and as an indifferent substitute for others, we are inclined to suspect its honesty; and as it is, moreover, a great enemy to the agriculturist, we do not scruple to include it among the "thorns and thistles" of the primæval curse. Yet, strange to say, as if to be both a memorial of the curse, and of the implied promise, that the industry of man should not be without effect in mitigating the consequences of that curse, the austere sloe has been converted by human skill and labour into the luscious plum, one of our most valued fruits. It is a well-known fact, that the thorns of several fruit trees, the Wild Pear for instance, disappear under cultivation; the variety of the Blackthorn, called the

vians learned the use of this bark by observing certain animals affected with intermittent complaints instinctively led to it: whilst others say that a Peruvian having an ague, was cured by happening to drink of a pool which, from some trees having fallen into it, tasted of Cinchona. It acts powerfully as an astringent and tonic; and as an antiseptic it is so efficacious as to preserve from decay not only animal solids but animal fluids when entirely detached from the living body. But its principal application is to the cure of intermittent fevers, when it rarely fails of success. The most valuable species of Cinchona are *C. micrantha* and *Condaminea.* Bark is not now administered in its native state so frequently as it formerly was, it having been discovered in 1820 that the active principle residing in it might be separated in the form of a crystallized salt by combining it chemically with sulphuric acid. In this state it is called *Sulphate of Quinin*, or simply *Quinin.* Its taste, like that of the bark itself, is excessively bitter: it is now generally employed as a substitute for bark, of which it possesses the medicinal virtues with this great advantage, that a few grains of the salt are equivalent to an ounce of the bark. The name *Quinin* is a corruption of *Quinquina*, another form of the word *Cinchona.*

Bullace tree,* is also entirely destitute of thorns, and produces edible fruit; while most of the kinds of plums cultivated in our gardens are referred by some eminent horticulturists† to the same origin. Every cultivator of dahlias or hearts-ease must

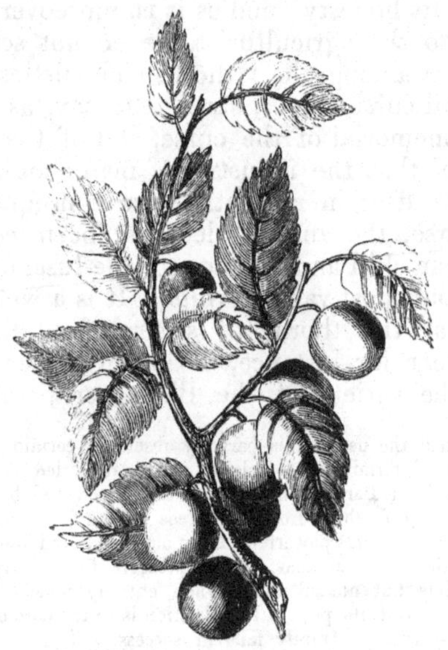

FRUIT AND FOLIAGE OF BULLACE TREE.

be aware that it is impossible to assign limits to the variations which these plants will undergo when subjected to the skilful treatment of the florist; and there is every reason, deduced both

* *Prunus insititia.* † Knight, Loudon, &c.

from theory and practice, why the same rule should be extended to fruit trees. In the Horticultural Society's Transactions, 274 distinct varieties of the plum actually in cultivation are enumerated, a number sufficiently great to admit of

MYROBALAN PLUM.

every possible gradation from the worthless sloe to the delicious green-gage. All these are referred by some horticulturists to another species, *Prunus domestica*, which, as its name would imply, is no longer found in a really wild state; and even when it is occasionally met with in hedges, approaches much more closely in character to the undoubtedly wild Bullace-tree, or Blackthorn, than it does to the garden varieties. The inference which we may safely draw from this fact is, that if the yellow *magnum-bonum* plum may be referred

MAGNUM-BONUM PLUM.

for its origin to the small black fruit of the "domesticated plum," as we find it in our hedges, we have at least equal reason for referring the latter to the sloe-tree.

For many of our best varieties of plum we are indebted to the French. First among these stands the Green-gage. It is known in France by several names: that of "Reine Claude" was given to it from its having been introduced into France by Queen Claude, wife of Francis I. During the Revolution, so great was the horror entertained against everything bearing the slightest allusion to royalty, that in order to retain its popularity it was obliged to change its name to "Prune citoyenne," *Citizen-plum*. It received its name Green-gage from the following circumstance. The Gage family, in the last century, procured from the monastery of Chartreuse at Paris, a collection of fruit-trees, the names of which were in every instance but one carefully attached to them. That of the Reine Claude, however, had been either omitted by the packer, or been rubbed off during the transit to England. The consequence was, that it stood without a name until it bore fruit, when the gardener very appropriately called it "Green-gage," in honour of the family who had introduced it. Since the revival of royalty in France, the Citizen-plum has recovered its ancient name, and "Reine-claudes" are now exported in large quantities.

The best prunes and French-plums come from Provence and the neighbourhood of Tours, the quality depending upon the sort of fruit used, and the care observed in the preparation. The commoner kinds are shaken from the tree and baked in an oven; but the finer sorts are gathered singly by the stems before sunrise, and laid, without touching one another, exposed to the sun and air several days before baking, great care being taken not to remove the delicate bloom with which they are covered.

Brignoles* are the dried fruit of a tree which grows principally near the town of the same name in Provence. They are peeled when fresh, and dried in the sun. When the moisture which they contained is entirely evaporated, the stones are taken out by hand, and the plums are pressed together in such a manner as to make them quite round. They are afterwards packed into small wooden boxes, ornamented with cut paper, and form an important article of revenue to the growers.

The Damascene, or Damson, takes its name from Damascus, where it grows in great quantities, and from whence it was brought into Italy about 114 B. C. It is used principally for preserves, and for making a kind of jelly called "Damson cheese."

Many kinds of plum were known to the Greeks and Romans; and Gerard had in his garden at Holborn, in 1597, "three-score sorts, all strange and rare."

For a fuller description of the garden-plums, I must refer my readers to works treating on horticulture.

The insects which prove most injurious to the Blackthorn and its varieties are, the brown and yellow-tailed moths, already noticed, and the following, a brief account of which may be acceptable, as displaying the wonderful instinct possessed by the meanest and most contemptible, (to use a common but an unconsidered expression,) of God's creatures.

The Copper-coloured Weevil (*Curculio* or *Rhyncites cupreus*) is a small beetle, less than two lines

* Corrupted into "*Prunellas.*"

in length, furnished with a long proboscis and hard wing cases, which are furrowed and metallic copper coloured. It is called in Germany *Pflaumenbohrer*,

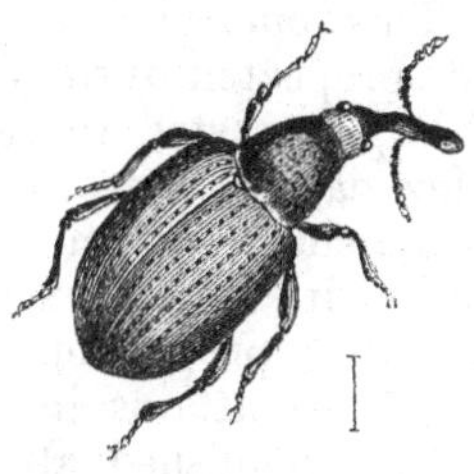

Curculio Rhyncites.
The line underneath denotes the natural size.

or *Plum-borer*, because it selects the plum for the reception of its eggs, and for the nourishment of the little larvæ proceeding from them. When the plums have attained the size of almonds, the weevil selects one in which to deposit an egg. As the larva when hatched feeds on the fruit, and as it can only be transformed into a beetle when buried in the ground, the parent-beetle is instructed to meet this difficulty, which she does most effectually. Having selected the plum which is to afford sustenance to one of her future progeny, she commences operations by sawing half-way through the stem of the fruit; and then, as if wearied with the sameness of her work, retires to the plum, and having prepared a nest by raising the skin and making a cavity underneath, deposits an egg at the entrance. She then turns down the skin, closing the orifice so effectually that not a drop of water can reach the egg, and continues her work on the stalk, either until the plum falls to the ground by its own weight, or it is left suspended by so fine a thread that the first violent wind completely separates it from the tree. This operation occupies two or three hours, and is repeated until all her eggs are laid, one only being intrusted to each plum. In the course of a few days, if the weather be favourable, the larva is hatched, and immediately begins to devour the pulp with which it is enveloped, enlarging its mansion in proportion to its growth. In five or six weeks it attains its full size, and having by this time demolished all of its habitation except the walls, eats its way out, and buries itself in the ground, where it forms a new habitation, and awaits its transformation. In the following spring it appears as a beetle, and in its turn begins the work of destruction. Immoderately wet or dry weather is equally prejudicial to its attaining the perfect state. If wet weather sets in, the plum soon becomes rotten and unfit for its use; and if dry, warm weather sets in, the half-grown plum and egg shrivel together. The excessive increase of these pernicious insects may be best checked by violently shaking the trees which appear to be prematurely shedding their fruit, and destroying all which fall to the ground.

Another insect which occasions the fall of the plum in its early stage is the Plum Saw-fly, *Tenthredo morio.* Seen at a little distance, it resembles a house-fly, but is strikingly distinguished by having four wings instead of two. The head and body are black, and the feet of a reddish yellow: it is not likely to be confounded with any other insect, as it is the only fly with feet of the same colour which appears when the plum is in blossom. As soon as the flower-buds of the plum-tree begin to expand, the fly cuts obliquely with its saws into the calyx of some of the larger kinds, and

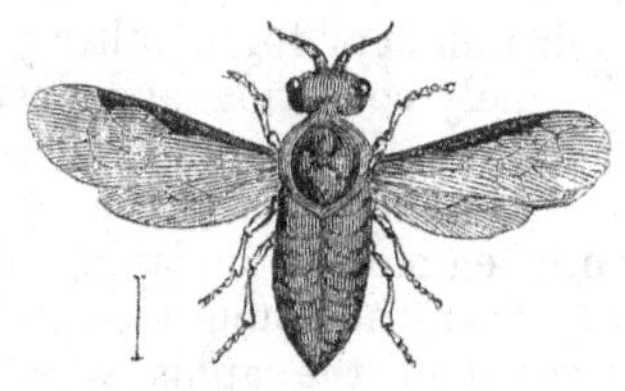

Tenthredo Morio—Plum Saw-Fly.
The line underneath denotes the natural size.

deposits its egg. In the course of a few days the larva is hatched, and immediately leaving its birth-place, where there is no suitable food for it, hastens to the minute plum growing near it, and fixes itself to the centre of the stone, which, being tender and milky, affords it the best nourishment. The plum meanwhile continues to grow, and the insect with it. In about six weeks, by which time the larva has attained its full size, the fruit and its destroyer fall together to the ground, when the latter buries itself in the earth, prepares its winter habitation, and emerges in the following spring a perfect insect.

So great is the mischief wrought by these flies, that they sometimes leave not a plum on the trees. In the year 1822, when their ravages were very extensive, Schmidberger had the plums on a Magnum-bonum tree, which promised to be very productive, counted: the number amounted to *eight thousand!* Only *three* plums arrived at perfection.

The weather was uncommonly favourable to them that year; they laid their eggs without interruption, and the larva found no difficulty in attaining its full development. A rainy season at the time when the Plum is in blossom is their greatest enemy. Their numbers for another year may be diminished by daily collecting and destroying the plums which fall to the ground. Their increase may also be materially checked by destroying the fly itself while employed in laying its eggs or sucking honey from the young blossoms.

A full account of the aphis which frequents many of our fruit-trees and garden plants will be given hereafter.

THE WILD CHERRY-TREE.

THE CHERRY.

CERASUS SYLVESTRIS.

CERASUS VULGARIS.

Natural order.—ROSACEÆ.

Class—ICOSANDRIA. *Order*—MONOGYNIA.

THE subject of the present memoir affords another eminent example, in addition to that recorded in the last chapter, of the beneficence of the Almighty in permitting man to control the course and operations of Nature, so as to render them, in a measure, subservient to his gratification and advantage. Human industry, we have seen, has converted the Thorn of the primæval curse into the fruitful Plum, and in the Cherry-tree we have another instance scarcely less remarkable; by dint of careful perseverance, a juiceless unpalatable seed becomes a delicious and nourishing fruit. The success which has attended the efforts of earlier cultivators ought, therefore, to supply us with a delightful incentive to industry, and, at the same time, a powerful motive to gratitude to our great Creator and Preserver. It were well if we never failed to raise our hearts in thankfulness to Him for the power which He has given us to employ for our own use and profit the "living things that move," which we have domesticated, and "the trees, in the which is the fruit of a tree yielding seed," which we have taught to be more productive and profitable. We should then bear in mind, that "the breath of life," which He breathed into our nostrils, is under Him our great instructor, and be prepared to submit ourselves in simple faith to His teaching, in matters where higher interests than things of the body are at stake.

Although no part of mighty Nature be
More stored with beauty, power, and mystery;
Yet to encourage human industry,
God has so order'd that no other part
Such space and such dominion leaves for art.
We nowhere art do so triumphant see,
As when it grafts or buds the tree.
In other things we count it to excel,
If it a docile scholar can appear
To Nature, and but imitate her will:
It over-rules, and is her master here.
It imitates her Maker's power divine,
And changes her sometimes, and sometimes does refine;
It does, like Grace, the fallen tree restore,
To its blest state of Paradise before.

COWLEY.

The Cherry-tree, though more familiarly known as a valued tenant of the orchard and garden, possesses undeniable claims to be considered a naturalized, if not a native, Forest Tree, resting its title both on its size and on the wildness of its haunts. It is not unfrequently met with in woods and hedges, and in the north of England is found on the mountains at an elevation of a thousand feet above the level of the sea. In congenial soils and situations it rises to the height of seventy or eighty feet, and in Scotland is planted for its timber. In some of the wilder parts of the same country it is as plentiful as the Birch, and propagates itself as freely.

In a picturesque point of view, its trunk and branches are light and graceful, but not sufficiently concealed by its scattered and somewhat scanty foliage. In early spring, however, the

FLOWER OF THE WILD CHERRY.

very deficiency of foliage renders more conspicuous its beautiful clusters of large flowers; while, in autumn, the bright crimson hue of its fading leaves irresistibly catches the eye, and imparts to the landscape a brilliancy which amply atones for any other defects. Amid mountainous scenery it is often particularly striking, contrasting exquisitely (especially when kindled into a brighter blaze by the straggling rays of the sun) with the dull grey of the rocks among which it has taken its station, and the rich brown of the river which it overhangs.

FRUIT OF THE WILD CHERRY.

There are several varieties of the tree even in the wild state; but modern botanists are of opinion that these may all be reduced to two species, the *Black* and *Red*-fruited.

It derives its name from Cerasus, (now Kerasoun,) a city of ancient Pontus, in Asia, whence it was brought by Lucullus, the Roman general, (B. C. 67,) at the close of the Mithridatic war. Lucullus thought this tree of so much importance, that, when he was granted a triumph, he placed it in the most conspicuous situation among the royal treasures which he had captured during the war; nor can there be any doubt, that, in permanent utility, it was the most valuable of his acquisitions. Some authors, however, are of opinion that the wild Cherry* was the same as the Cornel, which was indigenous in Italy at the time, but not cultivated as a fruit tree, and that Lucullus only introduced improved sorts. At all events, it does not appear to have been cultivated previously to the time of Lucullus, though afterwards it increased so rapidly, that, "in the course of a hundred and twenty years, it had reached even Britain. The Apronian cherries are red; the Lutatian black; the Cæcilian round. The Junian have a pleasant flavour; but are so tender, that they must be eaten under the tree, as they will not bear carriage. The best are those which in Campania are called 'Plinian;' in Belgium the 'Lusitanian' are considered the best. In the Rhine district grows another sort, which has a hue of black, red, and green, and never appears to be ripe. The Macedonian grow on a tree which rarely exceeds two cubits in height."†

* The fruit of this tree was subsequently called the Cornel-Cherry by some authors.

† Pliny's "Natural History," book xv. chapter 30.

According to the foregoing statement, the Cherry-tree was introduced into Britain before A. D. 53. The earliest mention of the fruit being exposed to sale by hawkers in London is in Henry the Fifth's reign, 1415. New sorts were introduced from Flanders, by Richard Haines, Henry the Eighth's fruiterer, and being planted in Kent, were called "Flanders" or "Kentish Cherries," of which Gerard (1597) says, "They have a better juice, but watery, cold, and moist." Philips says, "There is an account of a Cherry-orchard of thirty-two acres in Kent, which, in the year 1540, produced fruit that sold, in those early days, for 1000*l.*; which seems an enormous sum, as at that period good land is stated to have let at one shilling per acre." Evelyn tells us, that in his time (1662) an acre planted with Cherries, one hundred miles from London, had been let at 10*l.* During the Commonwealth (1649), the manor and mansion of Henrietta Maria, Queen of Charles I., at Wimbledon, in Surrey, were surveyed previously to being sold, and it appears that there were upwards of two hundred Cherry-trees in the gardens. Since that time the Cherry-tree has found universal admission into shrubberies, gardens, and orchards. Kent still continues the principal county for cherries; yet nowhere do they grow in greater luxuriance and beauty than on the banks of the Tamar, in Devonshire, where they freely "thrive into stately trees, beautiful with blossoms of a surprising whiteness, greatly relieving the sedulous bee, and attracting birds."*

* Evelyn's "Sylva."

In popular mythology the Cherry-tree is, for some unknown reason, associated with the cuckoo. In Germany, "the cuckoo never sings until he has thrice eaten his fill of cherries." In Yorkshire children were formerly, and perhaps still are, accustomed to sing round a Cherry-tree the following invocation:—

"Cuckoo, cherry-tree,*
Come down and tell me
How many years I have to live."

Each child then shook the tree, and the number of cherries which fell betokened the years of its future life.

At Hamburgh a feast is annually celebrated, called the "Feast of Cherries," in which troops of children parade the streets with green boughs, ornamented with cherries, to commemorate a triumph obtained in the following manner:—In 1432, the Hussites threatened the city of Hamburgh with immediate destruction, when one of the citizens, named Wolf, proposed that all the children in the city, from seven to fourteen years of age, should be clad in mourning, and sent as supplicants to the enemy. Procopius Nasus, chief of the Hussites, was so touched with this spectacle, that he received the young supplicants, regaled them with cherries and other fruits, and promised them to spare the city. The children returned crowned with leaves, shouting "Victory!" and holding boughs laden with cherries in their hands.

* A popular nursery rhyme begins with the same words.

The naturalized species of Cherry in Great Britain are the *Black* and *Red-fruited*, belonging to the genus *Prunus* of Linnæus, *Cerasus* of Jussieu.* *Prunus avium, Prunus Cerasus,* or *Cerasus sylvestris,* is the *Black-fruited* Cherry, which, in favourable situations, attains the dimensions of a tree. Its leaves are large, pointed, somewhat drooping, and slightly downy on the under side. The fruit is small, round, black when ripe, of an insipid bitterish flavour, and containing a stone which is very large in proportion to the size of the fruit. It is known in various districts by the name of Gean, (a corruption of *Guignes,*) Merries, (from *Mérisier,* said to be derived from *amêre,* bitter, and *cérise,* cherry,) Corone, or Coroun, (from *corone,* a crow, in allusion to its blackness,) Black-heart, &c.

"The growth of the Cherry, in its progress to maturity, is pyramidal; the branches springing from the stem at regular intervals, or at the commencement of each annual shoot: and as its spray is stiff, strong, and open, it does not yield to, but stoutly resists, the blast; it is, therefore, one of the few trees that can be advantageously planted as a nurse or subsidiary to the Oak, as it is neither apt to overtop or crush its neighbours by a rampant growth or wide-spreading head like the Wych-elm or Ash, or to hurt and injure them in winds and storms, as is constantly the case where trees with a more flexible or easily agitated spray are introduced. It has

* *Cerasus* may be distinguished from *Prunus*, by its leaves being conduplicate, or *folded* together in their young state, instead of being convolute or *rolled* together; and by the fruit being always destitute of the bloom which characterizes all the varieties of Plum.

also this further recommendation as a nurse to the Oak, that, although a quick-growing plant while young, and fulfilling the duty of a protector, it naturally yields to the tree it has fostered, after the first twenty or thirty years of its growth, and is afterwards content to vegetate beneath its shade. By producing suckers in abundance, it also furnishes a plantation with a profitable underwood, which may be cut once every five, six, ten, or more years, according to the purposes to which it is to be applied."*

Those botanists who are of opinion that Lucullus only introduced new kinds of Cherries into Europe, consider this species a native, and not without reason; for it grows freely and abundantly in Italy, Spain, Portugal, Turkey, Greece, Russia, the Mediterranean islands, Great Britain and Ireland, attaining a larger size in the north than in the south. Nevertheless, its general diffusion and apparent wildness of growth is not conclusive evidence in favour of its being considered a native of these countries. It has been remarked by M. le Conte, that in America, when Beech woods are cut down, they are speedily replaced by Cherry-trees. He accounts for this on the supposition, that birds, who eat the fruit with avidity, may have resorted to the woods for shelter, and there dropped the stones, which either lay dormant, or germinated and remained in a diminutive state until the Beeches were cut down, when they advanced rapidly, and finally became the principal occupants of the soil. Now, if the Cherry-tree has become thus thoroughly

* Selby's "British Forest Trees."

naturalized in America, into which there can be no doubt that it was introduced, there is very fair ground for the opinion that its extensive diffusion through Europe may be attributed to the same cause, and that the assertion of the older authors, that it is of Asiatic origin, is correct.

The second species, which, though often found in our woods and hedges, is not really wild in any part of Europe, is the Red-fruited Cherry. It is called by botanists *Prunus Cerasus*, or by those who assign the Plum and the Cherry to distinct genera, *Cerasus vulgaris*. To this species many of the best sorts of our garden Cherries are referred, including the Flemish and Kentish Cherries, Maydukes, (from Médoc, the province in France where the variety originated,) and many others. It is a much smaller tree than the last, from which it may be distinguished by its unpointed leaves, which are never downy beneath, and its red, acid, fruit.

In England, Cherries are to be considered rather as a luxury than as a staple article of food; but on the Continent, particularly in France, they are highly prized as supplying food to the poor; and a law was passed in that country, in 1669, commanding the preservation of all Cherry-trees in the royal forests. The consequence of this was that the forests became so full of fruit trees, that there was no longer room for the underwood; when they were all cut down, except such young ones as were included among the number of standard saplings required by the law to be left to secure a supply. This measure was a great calamity to the poor, who, during several months of the year, lived either directly or indirectly on the fruit. Soup made of Cherries, with a little bread and a little butter, was the common nourishment of the wood-cutters and charcoal-burners of the forest. Of late years the practice of planting Cherry-trees by the road-side has been extensively adopted in Germany; and one may now travel from Strasburg to Munich, a distance of two hundred and fifty miles, through an avenue of Cherries, interspersed with Walnuts, Plums, and Pears. By far the greater part of the first are ungrafted trees, which succeed in the poorest soil, and in the coldest and most elevated situations. A large portion of the tract of country which bears the name of Black Forest is an elevated, irregular surface, with no other wood than the Cherry-trees, which have been planted by the road-side.

Cherries are preserved in various ways. Sometimes they are simply dried in the sun, in which state they are much used for puddings; they are also preserved in brandy, or converted into marmalade, lozenges, &c. Fermented and distilled, they furnish the liqueurs called Ratafia, Kirschwasser, and Maraschino. Wine and vinegar are also made from them; and an oil is extracted from the kernels, which is used to give the flavour of bitter almonds to puddings, &c.; the leaves are also used for the same purpose.

From the bark of the Cherry-tree an elastic, but not very viscid gum exudes, which is said to have many of the properties of Gum-arabic.*

* Any excessive flow of gum is very injurious to the tree; and, indeed, in time proves fatal.

Hasselquist relates that more than a hundred men, during a siege, were kept alive for nearly two months, without any other sustenance than a little of this gum, taken sometimes into the mouth, and suffered gradually to dissolve.

"The timber is very valuable, being of a firm texture, red-coloured, close-grained, easily worked, and susceptible of a high polish. These qualities render it a desirable material to the cabinet-maker, and the furniture made of it is little, if at all inferior, both in respect to beauty and durability, to that of the plainer kinds of mahogany. In this country, where the wood just mentioned has in a great measure superseded all other kinds in our articles of furniture, and where the Cherry-tree has never been cultivated to any extent as a timber tree, it is rare to meet with specimens of furniture made of its wood; but in France, and other parts of the Continent where it abounds, it is extensively used for this and various other purposes, and is eagerly purchased by the cabinet-maker, the turner, and the musical instrument maker. Its value, however, is not restricted to the uses made of it by those artisans; it is equally applicable to out-of-door uses and general carpentry; and where exposure to the atmosphere, or the alternation of dryness and moisture is required, it is superior to most other timber we possess, and is only inferior to the best Oak, or its rival the Larch."*

When treated as coppice, it is very useful for hop-poles, props for vines, and hoops for casks.

* Selby.

The Turks have the tubes of their pipes, which are from four to seven feet long, made of Cherry stems.* Like the Ash, it burns very well as firewood in its green state; but if kept two or three years, and then used as fuel, it smoulders away like tinder, without producing much heat.

The double-flowered Cherry is a favourite ornament of our gardens and lawns in spring, when its numerous snow-white flowers present a beautiful appearance. Like many other double flowers, it produces no fruit; but the structure of its blossoms is particularly interesting to the physiological botanist, illustrating, better perhaps than any other plant, the fact that the seed-vessel, among other compound organs, is a metamorphosed or transformed leaf, altered in structure and functions, so as to perform offices in vegetable economy entirely different from those of the true leaf. In the double Cherry it appears to return to its primitive form; for in the centre of each flower is a minute leaf, exactly similar to those of the branches, notched and veined in the same manner, and even folded together like the young stem leaves. Other double flowers, beside those of the Cherry, occasionally present the same appearance, especially Roses; but in all these the phenomenon is an irregular mode of growth, whereas in the Cherry it is constant.

The Cherry is a favourite tree of the Woodpecker, who perforates its trunk for the sake of feeding on the larvæ of insects, and hollowing out his nest: but the remarks made at page 151 are equally applicable to the case of this tree.

* The best are made of the Mahaleb, or Perfumed, Cherry.

The Cherry-tree is not peculiarly liable to the attacks of insects. Its principal enemies are the Thrush and Blackbird, who annually claim a few Cherries in payment for their cheerful songs, and the pains which they bestow in clearing our gardens of snails and other vermin. One insect, however, of the class of spiders, (*Acarus telarius*,) known to gardeners as the Red Spider, occasionally does considerable injury. It has eight legs; its colour varies from yellowish to brown and reddish, and on each side of the back is a blackish spot. It is more frequent in the greenhouse than the open air. It spins a sort of web over the leaves, particularly on the under surface, and sucks the juice of the plants with its proboscis, completely enfeebling them, and stripping them of their leaves. The plants which it mostly attacks in the open air are the Kidney-bean, the Lime, and the Cherry.

Various remedies have been prescribed, which may easily be carried into effect in the greenhouse and hot-house; but in the open air, the only practicable preventive is to keep the tree in a healthy state, when the Spider will rarely touch it.

THE BIRD-CHERRY.

Cerasus Padus.

The Bird-Cherry in its wild state rarely attains the dimensions of a tree; but there are in existence cultivated specimens between thirty and forty feet high, and a foot or more in diameter. It is most worthy of attention for its copious long clusters of snow-white flowers, which are much smaller than those of the Cherry, and soon fade. The fruit, called also *Fowl-Cherry*, *Cluster-Cherry*, and in Scotland *Hag-Cherry*, is small and worthless. "Birds of several kinds soon devour this fruit, which is nauseous, and probably dangerous to mankind, though perhaps not of so deadly a quality as the essential oil, or distilled water of the leaves."* It is most abundant in the north of England and Scotland. In Gerard's time it grew wild in the woods of Kent, where it was used as a stock to graft Cherries on: and in Lancashire it was found in almost every hedge. The wood is much used in France by the cabinet-maker, but little known in this country; owing, among other causes, to the difficulty of obtaining it sufficiently large. The leaves are more frequently attacked by caterpillars than those of any other species of Cherry; hence, a writer in the *Agricultural Journal of Bavaria*

* English Flora.

recommends that from one to four young trees according to their size) should be planted at in-

BLOSSOM OF THE BIRD-CHERRY.

tervals of one or two hundred yards in orchards, when, he says, almost all the caterpillars and butterflies will resort to them. The appearance of the Bird-Cherry will be hideous, but the fruit trees will be safe.

Several other species of Cerasus are extensively cultivated in England as ornamental trees and shrubs, but none of them have any pretension to be admitted among British Trees. My notice

FRUIT OF THE BIRD-CHERRY.

of them therefore must be very brief. *Cerasus Laurocerasus*, the *Laurel-Cherry*, or, as it is now almost exclusively called, *Laurel*, was introduced into Europe from Trebizond, in Asia Minor, in 1576; consequently, it is a mistaken notion to identify it with the famed Laurel of the ancients. This error is the more frequent, from our having given to the true Laurel, *Laurus nobilis*, the name of Bay. Laurel leaves abound with prussic acid, and the water distilled from them is a most virulent poison. The custom of using them to flavour custards, puddings, &c., should therefore be strongly deprecated. Insects, the appearance of which is liable to be injured by immersion in spirits of wine, may readily be killed by being shut into a closed box with bruised leaves, the aroma from which speedily takes effect.

Cerasus lusitanica, or *Portugal Laurel,* is a native of the country from which it derives its name. It is not of rapid growth, but is a valuable acquisition to the shrubbery, from its elegance of form and hardy nature.

PORTUGAL LAUREL.

THE MOUNTAIN ASH.

THE MOUNTAIN ASH.

Pyrus aucuparia.

Natural order—Rosaceæ.

Class—Icosandria. *Order*—Pentagynia.

This universally admired tree chooses its dwelling, as its name would imply, in the wildest and most exposed situations, where, though impatient of being itself sheltered by any other kind of trees, it affords a friendly protection to grass and other plants which choose to grow beneath its shade. As long as it overtops its companions in the wood or mountain side, it is a vigorous and stately tree; but when it has attained its utmost height, and its more aspiring neighbours begin to screen it from its due share of air and light, it quietly retires from the contest, pines away in confinement, and suffers itself to be destroyed by the drip of the very trees that it formerly nursed and protected.

Hence we rarely meet with a full-grown Mountain Ash in a crowded forest of ancient trees. Where it has gained the vantage-ground of a broken rock partially covered with rich, light soil, or taken its stand in an open glade, amid plants of humbler growth, it attains a considerable size. Or, again, in an elevated situation, uncongenial to the rapid growth of its companions, but well suited to its own wild tastes and habits, it will continue to flourish for a century or more.

> The Mountain Ash
> No eye can overlook, when 'mid a grove
> Of yet unfaded trees she lifts her head,
> Deck'd with autumnal berries, that outshine
> Spring's richest blossoms ; and ye may have marked
> By a brook-side or solitary tarn,
> How she her station doth adorn : the pool
> Glows at her feet, and all the gloomy rocks
> Are brighten'd round her.
>
> WORDSWORTH.

The Mountain Ash is placed by most modern botanists in the same genus with the Apple and Pear, the fruit of which it resembles in conformation.* Others assign it a place with the Medlar, (*Mespilus*), or make it and the group with which it is connected a distinct genus (*Sorbus*). The name "aucuparia" (from *auceps*, a fowler) indicates the use to which its berries are applied by bird-catchers in France and Germany, who bait their traps with them as a certain lure for thrushes and fieldfares. Its popular names are very numerous: Mountain Ash, the commonest, is far from correct, as it belongs to an entirely different tribe from the Ash, which tree it resembles only in its leaves; Rowan, Roan, its common name in Scotland, and various other forms of the same word, occur in old authors. It is also called Quick-Beam, Wild or Fowler's Service-tree: "Service" appears to be a corruption of *Sorbus*, the ancient Latin name of an allied species, *Pyrus Sorbus*. Witchen, Wicken, Wiggen, &c. evidently

* The *Siberian Crab* (Pyrus baccata) produces fruit which may be considered as a connecting link between the berry of the Mountain Ash and the *Apple* of *Pyrus Malus*, the common Apple-tree.

bear allusion to the power it was once supposed to possess of counteracting witchcraft.

Lightfoot and Gilpin are both of opinion that the Mountain Ash was held in high estimation by the Druids. The former says, "It may to this day be observed to grow more frequently than any other tree in the neighbourhood of those druidical circles of stones so often seen in the north of Britain; and the superstitious still continue to retain a great veneration for it, which was undoubtedly handed down to them from early antiquity. They believe that any small part of this tree, carried about them, will prove a sovereign charm against all the dire effects of enchantment and witchcraft. Their cattle, also, as well as themselves, are supposed to be preserved by it from evil; for the dairy-maid will not forget to drive them from the shealings, or summer pastures, with a rod of the Rowan-tree, which she carefully lays up over the door of the sheal-boothby or summer-house, and drives them home again with the same. In Strathspey, they make, on the 1st of May, a hoop with the wood of this tree, and in the evening and morning cause the sheep and lambs to pass through it."

"In ancient days," says Gilpin, "when superstition held that place in society which dissipation and impiety now hold, the Mountain Ash was considered as an object of great veneration. Often, at this day, a stump of it is found in some old burying-place, or near the circle of a Druid temple, whose rites it formerly invested with its sacred shade." The custom of planting it in burying-grounds appears to have been retained after the introduction of Christianity; for Evelyn mentions, that, "in Wales, where this tree is reputed so sacred, there is not a churchyard without one of them planted in it, so, on a certain day in the year, every body religiously wears a cross made of the wood." In the Isle of Man, also, it is up to the present day invested by the superstitious with a sacred character. On Good Friday, when no iron of any kind must be put into the fire, and even the tongs are laid aside, lest any person should unfortunately forget the custom, and stir the fire with them, a stick of the Rowan-tree is used by way of substitute.*

The belief in the efficacy of the Mountain Ash, as a preservative against witchcraft, has led some commentators on Shakspeare to substitute, for the puzzling expression in "Macbeth," "Aroint thee, witch!" the words "A Roan-tree witch!" The passage being thus uttered, the mention of a tree so fatal to the power of the witch might naturally excite her acrimony against the person who applied the test. The authoress of "Sylvan Sketches" quotes a stanza from a very ancient song, which runs as follows:—

Their spells were vain; the boys returned,
To the queen in sorrowful mood,
Crying, that "witches have no power
Where there is Roan-tree wood."

In remote districts of England the superstition has not even yet died away. Waterton, in his "Essays on Natural History," relates an anecdote which fell under his personal observation, of a countryman in Yorkshire, who "cut a bundle of

* Train's "Historical Account of the Isle of Man, 1846."

Wiggin, and nailed the branches all up and down the cow-house," in order to counteract the effect produced on his cow by the "overlooking" of a supposed witch.

It is not a little singular, that, in like manner as we saw (page 182) similar superstitious practices holding in Ireland and the East with regard to the Hawthorn and a tree closely resembling it, so we find in India a tree bearing a strong resemblance to the Mountain Ash, to which the same superstition attaches.

Bishop Heber, in the 18th chapter of his "Indian Journal," gives the following account of this tree, and the superstition connected with it:— "As I returned home, I passed a fine tree of the Mimosa, with leaves at a little distance, so much resembling those of the Mountain Ash, that I was for a moment deceived, and asked if it did not bring fruit? They answered no; but it was a very noble tree, being called 'the Imperial tree' for its excellent properties, — that it slept all night,* and wakened and was alive all day, withdrawing its leaves if any one attempted to touch them. Above all, however, it was useful as a preservative against magic; a sprig worn in the turban or suspended over the bed was a perfect security against all spells, evil eye, &c., insomuch that the most formidable wizard would not, if he could help it, approach its shade. One, indeed, they said, who was very renowned for his

* Most plants of the Acacia tribe, which have compound leaves like the Ash, fold the leaflets together during the night, thus protecting their upper surfaces from the cold and damp. The same property resides in clover, and several other English plants of the same natural order.

power of killing plants, and drying up their sap with a look, had come to this very tree and gazed on it intently; 'but,' said the old man, who told me this, with an air of triumph, 'look as he might, he could do the tree no harm!' a fact of which I made no question. I was amused and surprised to find the superstition which, in England and Scotland, attaches to the Rowan-tree, here applied to a tree of nearly similar form. Which nation has been in this the imitator, or from what common centre are all these common notions derived?"

The Mountain Ash is found in a native state throughout the whole of Europe, and in several of the northern countries of Asia and North America. The parts of Great Britain where it attains its largest size are the western Highlands and the western coast of Scotland. On the hills of Cheshire and Derbyshire it does not often attain a great size: in such situations an entire tree, with roots, leaves, and flowers, is sometimes found not more than nine inches high. Ordinarily it grows very rapidly during the first five years of its existence, and at the age of twenty years forms a tree of the same number of feet with a single erect stem and a bushy head. The branches are smooth, and vary in colour from grey to purplish-brown. The buds, before their expansion in the beginning of April, are large and downy. The leaves consist of from seven to nine pairs of narrow, acute, notched leaflets, terminated by an odd one. These are somewhat downy underneath in their young state, but soon become quite smooth. The flowers are numerous, resembling in shape those of the Pear, but much smaller; in odour, those of the least fragrant varieties of Hawthorn. In early summer they are conspicuous from their number, and arrangement in large white clusters:

FLOWERS OF THE MOUNTAIN ASH.

when these are shed, the tree is still a pleasing object, from the brightness and elegant shape of its leaves. As autumn advances, it asserts its claims to be considered a fruit-tree, in appearance, if not for utility. Its flowers are then succeeded by numerous bunches of coral-red berries, which, until devoured by the Thrush and

FRUIT OF THE MOUNTAIN ASH.

Storm-cock, or scattered by the equinoctial gales, infallibly distinguish it from every other tenant either of the wood or the park. "In the Scottish Highlands, on some rocky mountain covered with dark Pines and waving Birch, which cast a solemn gloom over the lake below, a few Mountain Ashes joining in a clump, and mixing with them, have a fine effect. In summer the light green tint of their foliage, and in autumn the glowing berries which hang clustering upon them, contrast beautifully with the deeper green of the Pines; and if they are happily blended, and not in too large a proportion, they add some of the most picturesque furniture with which the sides of those rugged mountains are invested."*

A variety is cultivated which has yellow berries, and another with variegated leaves; but neither of these, as is the case with many other treasured rarities, has anything beyond its rarity to recommend it.

The berries, besides being applied to the use from which the tree derives its name, "Birdcatcher's Service," are eaten in the extreme north of Europe as fruit, though not, one would suppose, until every other kind of attainable fruit is exhausted, for they are intensely acid, and possess a peculiar flavour, which makes them very unpalatable. In seasons of scarcity, it is said that they are sometimes dried and ground into flour. "Some," says Evelyn, "highly commend the juice of the berries, which, fermenting of itself, if well preserved, makes an excellent drink against the spleen and scurvy. Ale and beer brewed with these berries when ripe, is an incomparable drink, familiar in Wales." A beverage resembling perry is still made from them in that country, and is much used by the poor. In Kamtschatka and in the Scottish Highlands an ardent spirit

* Gilpin.

is distilled from them, which is said to have a fine flavour.

As a timber-tree, the Mountain Ash does not attain a size which renders it available by the carpenter; but its wood, being fine-grained, hard, and susceptible of a high polish, is used for smaller manufactures, principally in turnery. As coppice it may be applied to most of the uses of Ash, Hazel, &c.; and the bark is employed by the tanner. In the days of archery, it ranked next to the Yew as a material for bows, and was considered sufficiently important to be mentioned in a statute of Henry VIII.

THE WHITE-BEAM.

PYRUS ARIA.

THE White-Beam* (or White-tree), though closely allied to the Mountain Ash, and consequently bearing a strong resemblance to it as to flower and fruit, is nevertheless very unlike it in general character and appearance. It is a native of the same countries, with the exception of North America, preferring chalky or limestone soils, where it frequently attains the height of thirty or forty feet. The trunk is straight and smooth, and the young shoots are covered with a white mealy down,

* "Beam," Saxon for "tree." So, in German, "Mehl-baum" means literally "Meal-tree," from the remarkably white and mealy appearance of the under side of its leaves.

as are also the under sides of the leaves, to such a degree as to give the tree its name. The flowers are larger than those of the Mountain Ash, and are succeeded by pale red berries, resembling in shape those of the Siberian Crab. Without being by any means common or well known, it occurs occasionally in various parts of England and Scotland. In the north of Devon I have seen it reaching a large size, and bearing abundance of fruit; but where the soil is not congenial, or the situation is confined, it scarcely merits the rank of a tree. The finest are said to grow near Blair, in Perthshire.

The fruit is used for the same purpose as that of the Roan-tree, and, if kept till it begins to decay, is somewhat more palatable, for in this state, like the Medlar, it loses a great deal of its austerity. It is eagerly devoured by birds, and on this account is in France protected by law, our neighbours having anticipated us in the discovery, that the hostility of birds against insects more than compensates in its effects for the occasional depredations which the former commit in our orchards and gardens. The wood of the White-Beam is very heavy and of a close texture, and is much used, especially on the Continent, for the cogs of wheels in machinery.

WILD SERVICE-TREE.

PYRUS TORMINALIS.

THIS species differs from the last in having its dark, glossy leaves lobed very like those of the Maple, whence it is sometimes called "Maple-Service." The fruit, which is brown and dotted when ripe, and not much larger than that of the Hawthorn, begins to decay when the frost has touched it, and is then agreeably acid and wholesome. Its geographical distribution is nearly the same with the White-Beam; but it is not found in Scotland or Ireland. It occurs occasionally in Cornwall as a hedge-bush, and in some other of the southern countries is said to attain the height of fifty feet; but it is nowhere common.

The "True Service-tree" (*Pyrus Sorbus*) is a doubtful native of Britain; but this is rarely met with even in a cultivated state, and requires no further mention.

The name "Service-tree" is often applied indiscriminately to all the above species of Pyrus, but belongs more particularly to the last.

THE PEAR.

PYRUS COMMUNIS.

AMONG the many industrial occupations which require the actual and continuous labour of the body, there is only one which is, from mere choice, extensively pursued by persons whose means and station in society exempt them from the necessity of manual labour; and that is the occupation of the gardener. We meet occasionally with persons in the higher walks of life, whose taste for mechanism takes them away in their leisure hours to active employment at the turning-lathe or the carpenter's bench; but this is so rarely the case, that any one who devotes himself to these pursuits must expect to incur the charge of singularity, or, at least, to be subject to remark among his neighbours. But, however exalted may be the rank of an individual who personally superintends, or engages in, the manual labour of rearing flowers or fruits, there are so many others who participate in the same tastes, that no one thinks it worth while to notice the fact. Nor does difference of station alter the case; for how often do we see the humblest artisan, when his daily toil is ended, and his frame, one would suppose, exhausted by previous exertion, stealing from the brief period of time allowed for refreshment and repose a large proportion to be devoted to his few square yards of garden ground! The very change of labour seems be his best restorative. He bends over his flowers and vegetables, and his work is now his recreation. Or, if debarred from renewing his toil in the open air, he expends his labour of love on a few cherished auriculas or carnations, and brings them to such perfection, that, while for their beauty and rarity they might vie with the costliest produce of his wealthy neighbour's hot-house, they impart to his homely cottage an air of elegance, which, if it were attached to articles of furniture or clothing, would be far above the station of the occupant; but, associated with this pursuit, tends to elevate the tone both of the dwelling and its inhabitants.

So common is this taste, that it excites little or no notice; and it seems so natural, that no one thinks of trying to account for its existence. It belongs to no particular age or climate, nor does it arise from any peculiar constitution of society; for even in so remote a country as China, and one the daily customs of whose inhabitants differ so widely from our own, the same fondness for the cultivation of fruits and flowers has existed for ages; and one of the most elegant and perfect poems of antiquity, the Georgics of Virgil, is devoted to the same and kindred subjects. The most remote history, sacred and profane, certifies the prevalence of the same taste, so that there can be little doubt of its existence in all ages and climes.

But whence did it originate? How comes it, that men of all nations take delight in an occupation so laborious as that of the cultivation of the soil? that civilised man, in all other cases so anxious to spare himself labour, considers this particular employment his privilege and relaxation? that the rude Indian, when torn from the natural garden of his forests and prairies, languishes and dies?

It is partly to be accounted for by the fact, that, in the operations of organized life, such as the unfolding of a flower, the ripening of a fruit, the withering of a leaf, the image of Nature presents itself most vividly to the soul; yet not wholly on this ground, for the argument would apply as strongly to the successive developments of animal life. We must therefore seek for a yet deeper reason, and that is supplied to us from the pages of Inspiration.

It was in a garden planted by the hand of God that man in his state of innocence first held intercourse with his Creator, and passed the only days of perfect happiness which man has spent on earth. After the fall, he "was sent forth to till the ground from whence he was taken," and, "in the sweat of his face," was sentenced to eat the produce of the soil. Yet, condemned to a life of toil as he thus was, it by no means follows that his labour was to be necessarily associated throughout with pain and suffering. A certain amount of anxiety and uncertainty was entailed on him, that he might not lose sight of his dependence on God, Who giveth the increase; but there was a reaping in joy, and a bringing home of sheaves with rejoicing, as well as a sowing in tears. In short, the labour of tilling the field has, by the Divine appointment, been from the beginning one which a faithful industry and a contented trustfulness have sweetened and sanctified. As mankind multiplied, and a changed system of society gave rise to new wants and new occupations, though they too were lightened and cheered by the solaces of hope and honesty, yet to his original employment man returns with an instinctive love, indicating its origin.

"God, the first garden made; the first city, Cain."

A warning against remission of industry was furnished by the fact, that to whatever degree plants may have been improved by cultivation, immediately on being neglected, they exhibit a strong tendency to return to their original wild state; and a further incitement to industry was afforded, by the discovery, that, although the mystery of creative power was placed out of reach, nay, out of sight, yet man was permitted, within limits, to bend the laws of nature to his will, and by skill and care, ingenuity and patience, to multiply new forms of existing plants to an indefinite extent.

By reference to these causes, principally, we are to account for the production of the countless varieties of flowers and fruits with which our gardens and orchards are filled,—varieties sometimes so unlike any known plants growing in a wild state, that it is hard to say from what stock they were originally derived. Pre-eminent among these, both for number and diversity of characters, stand the two trees of which we are now about to treat.

The Pear-tree, in its wild state, varies considerably in different countries, both in its mode of growth, and in the shape, size, and pubescence of its leaves. Some of these are probably distinct species, and inhabit most parts of Europe and Asia; but, as we have only to do with the British form of the tree, it is unnecessary to pursue this

FLOWER OF PEAR-TREE.

subject. It is found in most counties of England, growing in woods and hedges. Its outline, when it stands alone, is pyramidal: the branches are at first erect, then curved downwards and pendulous; in a truly wild state, thorny. The young leaves are slightly downy beneath, but, when mature, are quite smooth on both sides. When it is cultivated, the thorns on the branches disappear, as in the Plum. The flowers grow in clusters, and are large and of a pure white. The fruit is much smaller than that of any of the cultivated varieties, hard, austere, and unfit to eat; its only use is to mix with cultivated sorts in making perry. The wood was formerly sought after for wood-engraving, but is only adapted to coarse designs: it is also sometimes dyed black, in imitation of ebony.

For usefulness as a fruit tree, the Pear is rivalled only by, the Apple,—furnishing abundance of fruit, which is valuable in its fresh state, as well as for baking and preserving. Many sorts were well known to the Greeks and Romans; Pliny enumerates thirty-two. It was cultivated in England at a very early period. Chaucer makes mention of it; and in an account-book of Henry VIII. there are the following charges, among others:—

	£	*s.*	*d.*
"For medlars and wardens* . . .	0	3	4
Item, to a woman who gaff the Kyng peres .	0	0	2"

In Gerard's time, "threescore sundrie sorts of pears, and those exceeding good," were growing in one garden; and of late years so much attention has been paid to the multiplying of sorts, that the Horticultural Society's list for 1831 enumerates 677 named varieties.

The Pear-tree is long-lived, much more so in its cultivated than in its wild state; and its pro-

* "Wardens" were so called from their property of keeping: "peres" were probably some common kind of pear.

ductiveness increases with its age. Dr. Neile mentions a number of very ancient Pear-trees standing in the neighbourhood of Jedburgh Abbey, and in fields which are known to have been formerly the gardens of religious houses in Scotland which were destroyed at the Reformation. Such trees are, for the most part, in good health, and are abundant bearers; and, as some of them were probably planted when the abbeys were built, they must be from 500 to 600 years old.

The most remarkable Pear-tree in England stands on the glebe of the parish of Holme Lacy, in Herefordshire. When the branches of this tree, in its original state, became long and heavy, their extremities drooped till they reached the ground. They then took root; each branch became a new tree, and in its turn produced others in the same way. Eventually it extended itself until it covered more than an acre of ground, and would probably have reached much further if it had been suffered to do so. It is stated in the church register, that, "the great natural curiosity, the great Pear-tree upon the glebe, adjoining to the vicarage-house, produced this year (1776), fourteen hogsheads of perry, each hogshead containing one hundred gallons." Though now much reduced in size, it is still healthy and vigorous, and generally produces from two to five hogsheads. The liquor is not of a good quality, being very strong and heating. An idea of the superior size of this tree, when in its prime, over others of the same kind, may be formed from the fact, that in the same county, an acre of ground is usually planted with thirty trees, which, in a good soil, produce annually, when full grown, twenty gallons of perry each. So large a quantity as a hogshead from one tree is very unusual. The sorts principally used for making perry are such as have an austere juice.

The Pear is liable to be infested by several destructive insects, which prey either on the flower, fruit, leaf, or wood. The most remarkable of these is the Paradoxical Pear-fly (*Psilus Boscii*). This is a small black fly, scarcely a line long, furnished with a singular excrescence or horn rising from its back, which the insect keeps depressed close to the body, except when laying its eggs. These it deposits, in spring, in the opening blossoms of the Pear, to the number of six or seven. In a few days the larvæ escape from the egg and take refuge in the core of the young fruit, which, as if stimulated to unnatural exertion, increases rapidly in size, and soon outstrips the other pears, losing its bright green colour as it grows. It subsequently falls to the ground, when the larva escapes, buries itself in the ground, and remains there until it assumes the perfect state. The numbers of these insects may be sensibly diminished by collecting all the diseased fruits before they fall from the tree, and destroying them. Some entomologists are of opinion that this fly does not lay its eggs in the blossom of the Pear, but selects the fruit which is infested by the larvæ of another insect, the Black Gall-midge, (*Cecidomyia nigra,*) and deposits them in

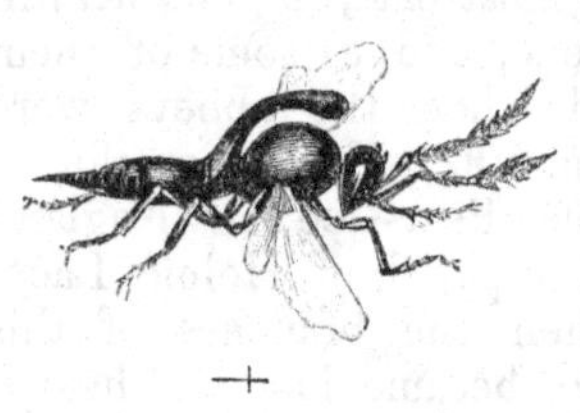

PARADOXICAL PEAR-FLY.

their bodies, just as the Ichneumon-fly lays its eggs in the larvæ of the *Cynips Quercus-folii* (page 42). In either case the remedy is the same. If a sharp frost sets in and destroys the blossom, these insects do not abound for some years.

A parasitic fungus (*Æcidium cancellatum,*) sometimes attacks the leaves of the Pear and commits great ravages. It first appears in the form of bunches of minute hairs on the veins of the leaves, always on the under side, and accompanied by a dingy-red spot above. When full grown, each spot consists of a number of bag-like excrescences, a quarter of an inch long, filled with seeds. Every leaf which is attacked dies; and, as the parasite when it does appear is very abundant,

ÆCIDIUM CANCELLATUM.

not only is the crop of fruit for the year deteriorated or totally destroyed, but the tree itself eventually perishes. Fortunately the disease is rare; as, when it appears in a garden, it defies all attempts to check it. A writer in the 'Gardener's Magazine,' (vol. ix. 333) states that on its first appearance it attacked only three trees in his garden; but the number gradually increased, until seventy, all in fact that stood in that part of the garden, had fallen victims.

THE APPLE.

Pyrus Malus.

The Apple-tree, being an undoubted native of Great Britain, demands to be noticed among our Forest Trees; though, from having been so long and so extensively cultivated, it is much better known as a tenant of the orchard than of the forest. Nevertheless, it is frequently to be met with in a perfectly wild state, possessing little or no value for its fruit, but forming in spring, with its rosy and fragrant buds, a beautiful ornament either to the woodland or the hedgerow.

It differs materially from the Pear-tree in shape, and is characterized by its crooked and knotty branches, which, if the tree is growing in an open space, spread equally on all sides, and give to it an irregularly hemispherical form. The leaves are generally wider in proportion to their length than those of the Pear, less pointed, and slightly downy beneath. The fruit may readily be distinguished by having its base, at the insertion of the stem, concave; that of the Pear being always convex. The branches are, both in the wild and cultivated states, destitute of thorns. It grows wild in most countries of Europe, and in western Asia, China, and Japan.

Improved varieties of the Apple appear to have been in cultivation from a very remote period. To the Greeks and Romans it was well known.

Mention of it occurs also in the Septuagint, as well as in the authorized version of the Holy Bible; but the fruit there alluded to is now thought, and with great propriety, to be the

Citron, which accords well with the description given in the Sacred Volume, and arrives at great perfection in Syria, whereas the Apple does not. The absurd legend, that the fruit of the forbidden tree was an apple, has probably given rise to the numerous superstitions respecting this tree, which appear under various disguises in the mythology of the Greeks* and Druids. The latter also looked on it with great veneration, from its being frequently clothed with Mistletoe. In certain parts of this country superstitious observances yet linger, such as drinking health to the trees on Christmas and Epiphany eves, saluting them by throwing roasted crabs or toast from the wassail-bowl to their roots, dancing and singing round them, lighting fires, &c. All these ceremonies are supposed to render the trees productive for the coming season.

I once had occasion to pass the night preceding Twelfth-Day at a lone farm-house on the borders of Dartmoor, in Devonshire, and was somewhat alarmed at hearing, very late at night, the repeated discharge of fire-arms in the immediate vicinity of the house. On my inquiring in the morning as to what was the cause of the unseasonable noise, I was told that the farm-men were firing at the Apple-trees in the orchard, in order that the trees might bear a good crop next season.

If these observances tended in the least degree to confer a benefit on the trees, they would not be mis-spent, for of all the fruit trees cultivated in

* The fable of the dragon which guarded the golden apples in the Garden of the Hesperides is probably derived from this source.

this country, the Apple is by far the most valuable, producing, with very little pains on the part of the proprietor, abundance of excellent fruit, fit either for the dessert, for dressing, or for making cider. To prove in what estimation it is held among gardeners, who resort to more sensible means for improving their trees than those above mentioned, it is only necessary to state that no less than 1400 named sorts, all differing from each other in shape, size, colour, flavour, or season of ripening, are enumerated in the Horticultural Society's Catalogue for 1831. All of these are cultivated in the Society's gardens, and new varieties are constantly being added.

The fruit of the wild Apple is called a crab, the sourness of which has passed into a proverb. The juice of crabs, called verjuice, is used to cure sprains and scalds, being often kept by good housewives in the country for that purpose. Isaac Walton, in his "Complete Angler," mentions it as being an ingredient in the rustic delicacy, syllabub. "When next you come this way, if you will but speak the word, I will make you a good syllabub of new verjuice, and then you may sit down in a hay-cock and eat it." The old-fashioned ointment called pomatum was made with the pulp of apples (*poma*), lard, and rose-water.

Though the Crab is the only Apple indigenous to Britain, several of the best sorts were first raised in this country. The Cornish Gilliflower is pronounced by Lindley the best eating apple; the Golden Pippin, so called from the small spots or pips that usually appear on the sides of these apples, is a native of Sussex; the Ribston

Pippin was raised at Ribston Park, Yorkshire, from a pippin brought from France. The original tree, which produced this last sort, was standing in 1831, and probably still remains. Philips, who published his poem, "Cider," in 1706, enumerates many sorts, some of which are still in cultivation; others have been superseded by more valuable kinds, or at least their names are rarely heard. Among these last is—

> "——John-Apple, whose wither'd rind, intrencht
> With many a furrow, aptly represents
> Decrepid age,"

and is no doubt the "Apple-John" of Shakspeare.

The Apple-tree is not remarkable for size or longevity, but is stated to be larger and more productive in North America than in Europe.

Darwin relates that in South America the Apple-tree attains great perfection. "The town of Valdivia," he says, "is situated on the low banks of a river, and is so completely buried in a wood of Apple-trees, that the streets are merely paths in an orchard. I have never seen any country where Apple-trees appeared to thrive so well as in this damp part of South America. On the borders of the roads there were many young trees evidently self-sown. In Chiloe the inhabitants possess a marvellously short method of making an orchard. At the lowest part of almost every branch, small conical, brown, wrinkled points project; these are always ready to change into roots, as may sometimes be seen where any mud has been accidentally splashed against the tree. A branch as thick as a man's thigh is chosen in early spring, and is cut off just beneath a group of these points, all the smaller branches are lopped off, and it is then placed about two feet deep in the ground. During the ensuing summer the stump throws out long shoots, and sometimes even bears fruit. I was shewn one which had produced as many as twenty-three apples, but this was thought very unusual. In the third season the stump is changed (as I have myself seen) into a well-wooded tree, loaded with fruit. An old man near Valdivia illustrated his motto, 'Necessity is the mother of invention,' by giving an account of the several useful things he manufactured from his apples. After making cider and likewise wine, he extracted from the refuse a white and purely flavoured spirit; by another process he procured a sweet treacle, or, as he called it, honey. His children and pigs seemed almost to live, during this season of the year, in his orchard."

It is somewhat singular that a very similar method of propagating Apple-trees is practised in so remote a country as China. The thick branch of a tree, when in full flower, is deprived of a ring of bark, and the place covered round with a lump of rich loam. This is kept moist by water, allowed to drip from a horn suspended above; and when the roots have pushed into the loam, which is usually the case when the fruit is nearly ripe, the branch is cut off and planted in a pot. Dwarf-trees, laden with fruit, are favourite ornaments among the Chinese. On the occasion of certain festivals, they are exposed on stands before the houses, along with grotesque figures of porcelain and pasteboard, which are made to perform a variety of absurd movements, by the agency of mice confined within them. Besides the Apple, the Orange and other kinds of fruit-trees are propagated in this way; and fine, that is, stunted and gnarled specimens fetch a high price. They are said to live from two to three hundred years, never much exceeding a foot in height, and producing annually from twenty to thirty large Apples. Several forest trees are treated in the same manner, particularly the Elm.

There formerly grew, on the eastern roof of the old Abbey Church of Romsey, Hampshire, a tree which regularly produced two kinds of Apples. How it came to grow in this place is not known. About a dozen years since, it being found that its roots were penetrating the stonework, and consequently were materially injuring the roof, it was destroyed.

The insects which select the Apple-tree for the food of themselves and their young are exceedingly numerous. Some of these, Loopers, (so called from their bringing forward their hindmost pair of feet in walking till they are close to the fore-feet, and so making with their bodies a bow or loop,) lay their eggs on the twigs nearest to the summit of the tree, to which they cement them so firmly, that no amount of rain washes them off, nor does the severest winter destroy their vitality. As soon as the flower and leaf-buds begin to expand, the young caterpillars burst from their shells, and commence the work of destruction, by eating their way into the buds, where they find both shelter and sustenance. They prefer at first the delicate food afforded by the tender unfolded petals and embryo fruit; but, as this is soon exhausted, extend their ravages to the leaves, the whole of the succulent part of which they demolish, and convert the withering skeletons and stems into habitations. When full grown, they let themselves down to the ground by a thread, which they spin from their mouths, and having buried themselves in the soil, there await their transformation.

One species of moth, the small Ermine Moth (*Tinea padella*), lays its eggs late in the summer on the small twigs, and cements them firmly to the tree, covering them with a strong gluten. The eggs are hatched the same year, but the grubs remain under cover during the winter. In the spring they issue forth with appetites sharpened by their long abstinence, and immediately eat their way into the substance of the young leaves, mining their course between the upper and under cuticle. As soon as they have outgrown the dimensions of their dwelling, they appear almost simultaneously on the outside of the leaves, and feed together in company under the protection of a common web, till the grub state of their existence is about to terminate, when they draw near together; each spins for itself a white cocoon, and is converted into a chrysalis.

ERMINE MOTH.

Many of these caterpillars become the prey of the Titmouse and various other small birds, which require a large supply of food for their young at the season when the caterpillars are most abundant. Ants also prey on them, and

Ichneumon flies contribute greatly towards checking their increase by laying their eggs in their bodies.

GRUBS OF ERMINE MOTH.

Several kinds of beetles, either in their larva or perfect state, attack the leaf, bark, or wood of the Apple.

> " Then the grub,
> Oft unobserved, invades the vital core,
> Pernicious tenant, and her secret cave
> Enlarges hourly, preying on the pulp
> Ceaseless."

This grub is produced either from the egg of the Codlin Moth (*Tortrix pomonana*) or, more rarely, from that of the Apple-Weevil (*Curculio Bacchus*). The moth lays one of its eggs in the eye of an apple when the fruit is well set. As soon as the egg is hatched, the caterpillar eats its way into the apple, avoiding the vital part or core, until nearly full grown; it then attacks the pips, and by the time that these are consumed, the apple falls to the ground, when the insect escapes, climbs up the stem of a neighbouring tree, and excavates for itself a dwelling in the bark, where it spins a white cocoon, and is converted into a pupa or chrysalis. The habits of the weevil-grub are nearly the same, except that it creeps into the ground to await its transformation.

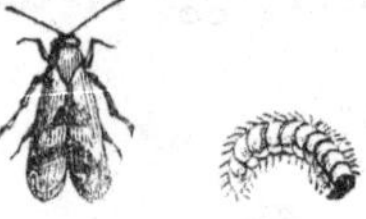

CODLIN MOTH.

The destructive insect called American blight, (for no other reason, one would suppose, than that it has been long the custom to ascribe the origin of most strange-looking things to the New World) is one of the greatest enemies of the Apple-tree. It is easily distinguished by its white cottony appendage, which is said to serve the double purpose of wafting the young insect through the air, when about to found a new colony, and of protecting it from the cold when established in its new dwelling. It injures the tree, and, if not checked, finally kills it, by sucking its juices through the bark. Many methods of destroying it have been suggested, among which, one of the simplest is to brush over every infected part with size. But even this remedy requires frequent repetition, as the insect infests even those parts of the tree which are beneath the ground. The subject is treated at length in the "Gardener's Magazine," vol. ix. p. 334.

The Apple-tree, both in its wild and cultivated state, is liable to be infested with the Mistletoe, which frequently does great injury.

In the west of England this parasite is but little known; but the Apple-trees, especially in the vicinity of the sea, are often so thickly invested with lichens, that the bark is scarcely to be dis-guished, except on the very young shoots. Most of them are of a pale ashen-grey or whitish tint; one, however, which occurs but rarely in the eastern counties, *Borrera flavicans,* is very conspicuous for its tangled golden tufts, which in winter, when the tree is divested of foliage, are very ornamental.

I must not omit to mention that the Mistletoe Thrush, or Storm-cock, which at most seasons is one of our wildest birds, in spring deserts its favourite tree, the Mountain Ash, and resorts to the neighbourhood of human dwellings. There it selects, as a fit place for rearing its young, an Apple-tree close to the house, choosing the angle between the trunk and one of the principal branches. It builds its nest of materials which closely resemble the bark of the tree, and, though exceedingly shy at other seasons, now sits so closely, that one may advance to within a few yards of the nest without being noticed. The beautiful copper-coloured Chaffinch also prefers to build her elegant nest among the twigs of the Apple-tree, and decorates it in the neatest manner with the lichens which infest the tree she has selected.

THE PURLEY BEECHES.

THE BEECH.

FAGUS SYLVATICA.

Natural order—AMENTACEÆ.

Class—MONŒCIA. *Order*—POLYANDRIA.

THE Beech, though one of our most abundant forest trees, growing spontaneously in the wildest parts of many of the counties of England, perfecting its seed freely, and sustaining a vigorous growth, (which proves that the soil and climate of the country are perfectly congenial to it,) is nevertheless declared by many writers to be a doubtful native. This opinion they justify on the ground that Julius Cæsar, in his account of his invasion of Britain, states that "timber of every kind which is found in Gaul also grows in Britain, except the Beech and the Silver Fir."* The fact is, that by far too much importance is attached to this passage. Cæsar penetrated but a very little way into Britain, staid there but a short time, and rarely ventured to any great distance from the camp; consequently he saw very little of the country. There can be no doubt, however, that he was anxious to convey to his countrymen as favourable an impression as possible of his achievements; so that, the suc-

* "Materia cujusque generis, ut in Gallia est, præter Fagum et Abietem." (*Cæsar de Bel. Gal.*)

cess of his military operations being slight, he would very willingly have them infer, from the minuteness with which he particularised the produce of the island, that he had penetrated far into the country, but had met with no adventures worth recording. This seems the readiest way of meeting the difficulty. Other writers suggest that some other tree than the Beech may be identical with the Fagus of Cæsar, and have endeavoured to show that he meant the Chestnut. But that this opinion is erroneous, will appear from the following consideration.

The Roman poets make frequent mention of the tree (*Fagus*) which Cæsar declares to be not a native of Britain.* They describe it as being lofty, furnished with wide-spreading branches, casting a dense shade, loving the hill side, attaining a great age, and furnished with so smooth a bark, that rustics selected it to carve their names on, and even for the reception of their poetical effusions.† Virgil states that it was

* The φηγὸς (*phegos*) of Theophrastus does not appear to be the same as the Fagus of the Romans, though both names have the same etymology, from φάγω (*phago*), to eat. Our Beech is most probably the tree which that author calls αἰγίλωψ (*ægilops*), and describes as "a mast-bearing tree, furnished with a very straight trunk, very lofty, having a smoother bark than any of the other mast-bearing trees, and growing but sparingly in enclosed country." (*Theophrastus de Plantis*, lib. ii.) The φηγὸς of Theoprastus was probably the *Æsculus* of the Romans.

† Among the many anecdotes connected with the history of printing, which have come down to us, that related by Hadrian Junius deserves to be noticed in this place. About the year 1441, Lawrence Koster, a citizen of Haarlem, "walking in a suburban grove, began first to fashion Beech-bark into letters, which being impressed upon paper, reversed in the manner of a seal, produced one verse, then another, as his fancy pleased, to be for copies to the children of his son-in-law." This hint he subsequently improved upon,

grafted on the Chestnut, and that its wood was converted into bowls, a use which is alluded to by other poets. No other tree with which we are acquainted accords with this description. But this is not all, for Pliny, the Latin naturalist, gives an accurate description of the Fagus, which cannot fail to identify it with our Beech. "Of the various kinds of mast, that of the Fagus is the sweetest, on which Cornelius Alexander says, that some men, who were besieged in the town of Chios, lived for some time. It resembles a nut, and is enclosed in a triangular rind. The leaf is thin and exceedingly smooth, shaped like the Poplar, decaying after it has fallen to the ground long before any of the other mast-bearing trees. The mast is much eaten by mice, which abound at the season of its ripening; it also entices dormice, and is much sought after by thrushes. Hogs fattened on it are lively, and their flesh is digestible, light, and wholesome. The bark is used for making baskets and panniers, but the timber is not durable."

The above description, though wanting the precision of modern science, is sufficiently conclusive that the Fagus of Italy is the Beech of Great Britain, for the account is not true of any other known tree. The only statement which demands further notice is that of Virgil, that the Beech is often grafted on the Castanea or

and finally invented blocks of lead and tin, and printed books. Among his workmen was John Faust, who, having been initiated in the art, although sworn to secrecy, decamped, carrying with him his master's stock in trade, and set up as a printer on his own account at Mayence. I should add, that, although many literary men have credited this account, it bears, on close examination, internal evidence of being a fabrication, either of Hadrian or his informant.

Chestnut. This assertion has appeared so strange and unaccountable to commentators, that some have got rid of the difficulty at once by supposing that the passage is corrupt, and that Virgil meant to say, "the Chestnut is often grafted on the Beech:" others have jumped to an equally unwarrantable conclusion, that the Beech was called by the Romans "Castanea," and the Chestnut "Fagus;" and that, accordingly, Cæsar asserted that the *Chestnut* did not grow in Britain. This ingenious explanation is so satisfactory, that it might be adopted at once, if sufficient evidence of the fact could be adduced. But this is not the case, for Pliny's description of the Castanea agrees as exactly with the Chestnut, as that of the Fagus does with the Beech. "The fruit of the Castanea," he says, "we call also a nut, though it approaches nearer in character to mast. It is protected by a case beset with strong prickles. It is strange that we hold as of no value a fruit which Nature has so carefully guarded from injury. As many as three nuts frequently grow together in one case. The proper rind of the nut is tough, and within this is a thin skin closely attached to the substance of the nut, as in the walnut, which, unless it be removed, spoils the flavour of the fruit. The best way of preparing them for food is by roasting. They are sometimes ground into meal, which is converted by women into a wretched substitute for bread, and eaten during their religious fasts."

From a comparison of these passages, it will plainly appear that the tree which we call Beech was undoubtedly the Fagus of the Romans, and the Chestnut, Castanea. Nor will there be any difficulty in discovering the propriety of grafting the Beech on the Chestnut, the oily though smaller nut of the former being considered by the ancients much more valuable than the farinaceous nut of the latter.

On the whole, therefore, the readiest solution of the difficulty is, that Cæsar did not penetrate into any part of the island where Beeches were abundant, and that the woods, to which he tells us that the Britons retired to escape from their invaders, were composed of trees which admitted a more luxuriant growth of underwood than this unsociable tree ever allows.

Loudon states that it is a native of the temperate parts of Europe, from the south of Norway to the Mediterranean Sea, and from England to Constantinople. It is also found in Palestine, Asia Minor, and other parts of Asia. In Switzerland it occupies the south sides of the mountains which have their north sides clothed with the Silver Fir. In England it grows most luxuriantly and in the greatest abundance in the chalk districts, forming extensive forests of great magnificence and beauty. It is not indigenous to Scotland or Ireland. It is the national tree of Denmark, and in the neighbourhood of Elsinore flourishes in superlative vigour.

In North America, a species very similar to the Beech of Europe forms extensive woods in the middle and western states. In South America its place is supplied by other species, *Fagus betuloïdes* and *F. antarctica,* though only in the extreme south. In Tierra del Fuego, the former frequently measures as much as thirteen feet in circumference. Captain King mentions one which was seven feet in diameter at seventeen feet above the roots. Darwin thus describes his attempt to penetrate a Beech forest in that country:—"Finding it nearly hopeless to push my way through the wood, I followed the course of a mountain torrent. At first from the waterfalls and number of dead trees, I could hardly crawl along; but the bed of the stream soon became a little more open, from the floods having swept the sides. I continued slowly to advance for an hour along the broken and rocky banks, and was amply repaid by the grandeur of the scene. The gloomy depth of the ravine well accorded with the universal signs of violence. On every side were lying irregular masses of rock and torn-up trees: other trees, though still erect, were decayed to the heart, and ready to fall. The entangled mass of the thriving and the fallen reminded me of the forests within the tropics: yet there was a difference; for in these still solitudes Death, instead of Life, seemed the predominant spirit. I followed a watercourse till I came to a spot where a great slip had cleared a straight space down the mountain side. By this road I ascended to a considerable elevation, and obtained a good view of the surrounding woods. The trees all belong to one kind, the *Fagus betuloïdes*, (the Birch-like Beech,) for the number of the other species of Fagus and of the Winter's Bark* is quite inconsiderable. This Beech keeps its leaves throughout the year; but its foliage is of a peculiar brownish-green co-

* This tree, *Drimys Winteri*, is closely allied to the genus Magnolia, and furnishes the aromatic Winter's Bark, which is remarkable for its resemblance to that of Cinnamon.

lour, with a tinge of yellow. As the whole landscape is thus coloured, it has a sombre, dull appearance, nor is it often enlivened by the rays of the sun." On another occasion, when he accompanied the commander of the expedition to explore the Beagle Channel, "the view," he says, "was very remarkable. Looking towards either hand, no object intercepted the vanishing points of this long canal between the mountains. The circumstance of its being an arm of the sea was rendered very evident, by several huge whales spouting in various directions. On one occasion I saw two of these monsters, probably male and female, slowly swimming one after the other, within less than a stone's throw of the shore over which the Beech-tree extended its branches."*

A species of Beech which grows at Van Diemen's Land attains a height much greater than that of any European tree.

The Beech was particularly admired by the ancients, who luxuriated in the lofty canopy afforded by its dense foliage. In modern times, its claims to the possession of picturesque beauty have been disputed on high authority, for while Gilbert White speaks most warmly in its praise, Gilpin expresses a very different opinion. The former, in describing the parish of Selborne, says, "The high part to the south-west consists of a vast hill of chalk, rising 300 feet above the village, and is divided into a sheep down, the High Wood, and a long hanging wood called the Hanger. The covert of this eminence is altogether Beech, the most lovely of all forest

* Journal of Researches into the Natural History and Geology of the Countries visited by H. M. S. Beagle.

trees, whether we consider its smooth rind or bark, its glossy foliage or graceful pendulous boughs." Gilpin, after pointing out the defects of the timber of this tree, proceeds to say, "In

VILLAGE OF SELBORNE.

point of picturesque beauty, I am not inclined to rank the Beech much higher than in point of utility. Its skeleton, compared with that of the Oak, the Ash, and the Elm, is very deficient. Its trunk, we allow, is often highly picturesque; it is studded with bold knobs and projections, and has sometimes a sort of irregular fluting about it, which is very characteristic.

It has another peculiarity, also, which is sometimes pleasing,—that of a number of stems arising from the root. The bark, too, often wears a pleasant hue. It is naturally of a dingy olive; but it is always overspread, in patches, with a variety of mosses and lichens, which are commonly of a lighter tint in the upper parts, and of a deep velvet-green towards the root. Its smoothness also contrasts agreeably with these rougher appendages. But having praised the trunk, we can praise no other part of the skeleton. The branches are fantastically wreathed and disproportioned, turning awkwardly among each other, and running often into long unvaried lines, without any of that strength and firmness which we admire in the Oak, or of that easy simplicity which pleases in the Ash; in short, we rarely see a Beech well ramified. In full leaf it is equally unpleasing; it has the appearance of an overgrown bush. This bushiness gives a great heaviness to the tree, which is always a deformity: what lightness it has, disgusts. You will sometimes see a light branch issuing from a heavy mass; and though such pendent branches are often beautiful in themselves, they are seldom in harmony with the tree. They distinguish, however, its character, which will be seen best by comparing it with the Elm. The Elm forms a rounder, the Beech a more pointed, foliage; but the former is always in harmony with itself.

"Sometimes, however, we see, in Beeches of happy composition, the foliage falling in large flocks or layers elegantly determined; between which the shadows have a very forcible effect, especially when the tree is strongly illumined. On the whole, however, the massy, full-grown, luxuriant Beech is rather a displeasing tree. It is made up of littlenesses, seldom exhibiting those tufted cups, or hollow dark recesses, which dispart the several grand branches of the more beautiful kinds of trees. Contrary to the general nature of trees, the Beech is most pleasing in its juvenile state, as it has not yet acquired that heaviness which is its most faulty distinction. A light, airy, young Beech, with its spiry branches hanging, as I have just described them, in easy forms, is often beautiful. I have seen, also, the forest Beech, in a dry, hungry soil, preserve the lightness of youth in the maturity of age.

"After all, however, we mean not to repudiate even the heavy, luxuriant Beech in picturesque composition. It has sometimes its beauty, and oftener its use. In distance, it preserves the depth of the forest; and even on the spot, in contrast, it is frequently a choice accompaniment. In the corner of a landscape, when we want a thick heavy tree, or part of one at least, which is often necessary, nothing answers our purpose like the Beech. But at present we are not considering the Beech in composition, but only as an individual; and in this light it is which we chiefly conceive it as an object of disapprobation."

Now it is very clear that the two authors whom I have just quoted, at the time when they described the tree, were actuated by very different feelings, White approaches it as a genuine lover of nature, with a vision quick to discover, and predisposed to admire, all that is beautiful in its form and colouring, admirable in its structure, or impressive in its proportions—every reality of nature, in short, which might present itself to him in its perfect state. It does not occur to him to consider what combinations of the Beech with other objects would make a beautiful picture, or how the painter would manage the lines formed by the branches in transferring them to his canvass. He discovers no "awkwardness" in the intertwining of the limbs, and feels no "disgust" at the lightness of the spray: the former is to him the natural characteristic of the tree; and he is well pleased to look up and admire the delicate twigs with their scattered leaves painted on the curtain of the heavens. He sees things as they exist in nature, and is not for the time aware that to the artist they have another, independent existence: he is, as a naturalist, unconscious that what is beautiful in nature is not of necessity picturesque in art. Gilpin, however, though he might, if he chose, divest himself of the feelings of the painter, and then admire all that seemed admirable to the other, cannot do so without an effort. Carried away by the same feelings, he pronounces also a very harsh judgment on the Hawthorn.*

Sir T. D. Lauder, remarking on Gilpin's strictures on the Beech, says, with great propriety, "It must be observed here, that this is one of the instances in which the author's love for the art of representing the objects of nature with the pencil, and his associations with the pleasures of that art, have very much led him away. We are disposed to go along with him in a great measure, so far as we, like him, draw our associa-

* See page 199.

tions with this tree from the same source. But we conceive we have much the advantage of him, in being able to indulge in the pleasure arising from the contemplation of a noble Beech as one of the most magnificent objects of God's fair creation. Some of the very circumstances which render it unpicturesque, or, in other words, which render it an unmanageable subject of art, highly contribute to render it beautiful. The glazed surface of the leaf, which brightly reflects the sun's rays, and the gentle emotions of light, if we may venture so to express ourselves, which sometimes steal over the surface of its foliage with the breathing of the balmy breeze, although difficult, or rather almost impossible, to be represented by the artist, are accidents which are productive of very pleasing ideas in the mind of the feeling observer of nature."

On the whole, therefore, without going so far as to assert that none of the objections alleged against the Beech by our great authority on forest scenery are tenable, I may say with safety, that, in spite of them all, the Beech is a noble tree in nature—beautiful, as delineated by the hand of the Creator, however difficult it may be for the painter to represent it with the pencil in such a way as to produce a pleasing effect on the mind. And it is a tree which has many points of interest about it at all seasons of the year. Enter a grove of Beeches on a bright day in mid-winter; the mind is immediately engaged in meditating on the still solemnity that reigns around. Look where you will, Nature is in a state of deep repose, if not of suspended animation: there is as little semblance of growing life as in the cloisters of a cathedral. The ground is bare of everything save withered leaves, and dead twigs, and wrinkled husks; every herb, if any ever grew here, has hidden itself under the brown covering of the earth, as if afraid to show signs of life in that universal solitude. As far as the eye can reach, on all sides extends an irregular succession of lofty fluted columns, which seem to have been chiselled to their existing proportions; for nowhere is there to be detected a single rugged trunk indicative of expansive growth, nor one to which the mantling Ivy imparts a borrowed semblance of vitality. The very lichens which chequer their smooth barks seem to be monumental, rather than endowed with life. Overhead, the long wavy boughs are intersecting each other at every possible angle, but all stark and rigid. The wiry twigs, which form a network over the whole, are apparently striving to escape from the solemn influence which reigns below. Yet there is no gloom here, for the sun, as if aware that this is the only season at which his rays can penetrate these recesses, makes up in brightness for what he wants of heat. And, if we look a little more closely, we shall discover that, though Nature is asleep, her vital functions have only withdrawn themselves from sight: mysterious operations are still going on, of which, though we cannot now comprehend them, we shall in a few months have no difficulty in discovering the results. Examine one of the long and sharp buds with which every branch is so plentifully furnished, and, although we may be unable to account for the apparent suspension of life in deciduous* trees, or to discover what operations are being carried on in the silent laboratory of Nature, we shall have no difficulty in discovering that the providence of God is watch-

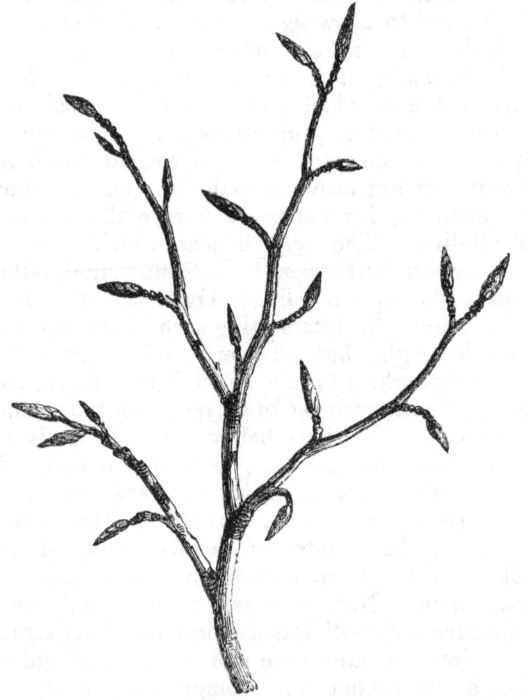

ing over every bud, and doing for it whatever is necessary, in order that it may, at the return of spring, be converted into a leafy shoot. Wrap-

* Deciduous trees are those which shed their leaves at the approach of winter.

ped up in a mantle of silk and waterproof scales, the tender nursling is protected against wind, and rain, and cold, and is provided with all that it needs in order to maintain a healthy existence, but not with that, whatever it may be, which could stimulate it to throw off its integuments and come forward into the light of Heaven before the time assigned by its Maker. Examine again the younger trees on the skirts of the grove. They are still clothed with the shrivelled foliage of the preceding summer. One would imagine that, exposed to the autumnal blasts as they have been, they would be the first to shed their leaves. But no! after these had fulfilled their office as living organs, another remained to be performed, and they must stay where they are until thrust off in the spring by the expanding buds. We know not what their office is; perhaps it is to protect the embryo leaves of the coming year, while the tree is yet young and tender: but even though we may be wrong in our surmise, the error cannot be an important one, if it has led us to meditate faithfully on the watchful superintendence which God exercises over all the works of His creation.

We may often see, on the bole of a Beech, scattered excrescences called "knurs," varying in size from a pea to a large marble. They may be separated from the tree by a smart blow with a stick, and are found to be composed of a solid ball of wood, surrounded by a layer of bark like that of the rest of the tree. The woody part is completely imbedded in bark, from which they would appear to have been deposited, thus confirming the well-known fact that such is universally the origin of woody fibre. Whether they originated from the puncture of the bark by an insect, or from what other cause, is not known. If planted in the earth, they will grow:* but I am not aware whether they ever shoot forth while attached to the tree.

Come again to this spot,

> "when rosy-footed May
> Steals blushing on."

The delicate leaves, with their glossy silk fringe, now so carefully wrapped up in the bud of from twenty to thirty membranous scales, will then be shewing their vivid green on the lower branches, the bud scales as yet loosely clinging to their base. In a few days more the verdure creeps up the whole tree, gradually deepening in hue, and assuming a brighter polish. The silken fringe, which was so conspicuous when the leaf emerged from its winter's covering, becomes more scattered as the leaf increases in size, the latter being slightly notched, and having the veins beneath downy. But we shall look in vain for a carpet of herbage beneath its shade. Here and there a sickly Holly has resisted the malignant influence of its drip, or a tangled bed of Periwinkle† has established itself, and grows on luxuriantly, unaffected by the prevailing cause of sterility: but, with these exceptions, the Beech has appropriated the whole of the soil. Where it has obtained the sway, it suffers no other ver-

* I state this on the authority of Mr. J. Sidebotham, an intelligent botanist of Manchester.

† *Vinca minor.*

dure to exist. Consequently, the ground, covered with decaying leaves at all seasons of the year, always presents the same appearance. As summer advances, a few Orchideous plants* may be detected here and there, but not sufficiently numerous or striking in appearance to alter the character of the scene.

FOLIAGE AND FLOWERS OF THE BEECH.

By the time that the foliage is fully developed, the flowers also have made their appearance. These are of two kinds: the barren, which are of a brown hue, three or four together in round

* *Gymnadenia bifolia, Listera nidus-avis,* &c.

drooping heads;* the fertile flowers are solitary and on stouter stalks. The first soon wither and drop off; the latter produce seed-vessels, which are covered with blunt prickles, and, as they ripen, open in four valves, disclosing two sharply triangular pointed nuts. It is when seen in the full luxuriance of its summer foliage that the Beech is most admired: at this season it is, if a solitary tree, a mass of shining deep green, from the ground to its summit; and the lover of nature, who has taken refuge in a grove of Beeches from the sultry heat of a cloudless summer's day, will not fail to experience that inexplicable feeling of sadness, mingled with longing, which the contemplation of Nature's greater works always excites.

"Under the broad Beech-tree" honest old Isaac Walton loved to sit, "viewing the silver streams glide silently towards their centre, the tempestuous sea;" and, as he thus sat, "these and other sights so fully possest his soul with content, that he thought, as the poet has happily exprest it,

'I was for that time lifted above earth,
And possest joys not promised in my birth.'"

Fletcher chooses the same retreat for the humble and contented hero of one of his lays:—

"No empty hopes, no courtly fears him fright,
No begging wants his middle fortune bite;
But sweet content exiles both misery and spite.

* These, after they have fallen from the tree, are sometimes carefully collected by gardeners, dried, and preserved for packing fruit. They are as soft as cotton, and do not communicate any kind of scent to the fruit.

His certain life, that never can deceive him,
Is full of thousand sweets, and rich content:
The smooth-leaved Beeches in the field receive him
With coolest shade, till noon-tide's heat be spent.
His life is neither tost in boisterous seas,
Or the vexatious world, or lost in slothful ease;
Pleased and full blest he lives, when he his God can please."

The grateful coolness perceptible in woods of Beech, and indeed all trees which cast a deep shade, is produced by the combined influence of three several causes, each depending on a distinct physical fact. The first, which may be said to be purely mechanical, is so plain, that every one must be acquainted with it: the direct rays of the sun are intercepted by the foliage, and are thus prevented from heating the ground and the air. But how comes it that the same effect is not produced by any other kind of shelter, the roof of a house or tent for instance? The latter intercepts the sun's rays more completely, perhaps, than the leafy shelter of the grove, for, however thick the foliage may be, a few straggling rays contrive to force a passage through. Yet the fresh coolness of an over-arching Beech-grove is as different as possible from the mere shelter afforded by any artificial roof.

The reason is, that, in the former case, two natural operations are simultaneously going on, which have the effect of cooling the air in the vicinity of the foliage, and of preventing the covering itself from becoming heated; whereas, in the case of the artificial roof, only one of these causes operates at all, and that in a very limited degree. These are *radiation of heat* and *evaporation*. All bodies possess the property of parting with their heat, which is constantly proceeding in all directions, and in straight lines from every part of their surface, the quantity of heat thus lost or *radiated* being proportionate to the extent of surface of the body. Now it is evident that the quantity of surface contained by the trunk, branches, and leaves of a tree must be many times greater than that of the ground which the tree covers; consequently, the diminution of heat must be greater in the same degree. For this reason alone, we should expect to find the leaves of a tree much cooler than the bare ground, supposing that both were alike submitted to the rays of the sun. To this cause must be added another. Every plant, during its state of active growth, that is, as long as it is in leaf, and is exposed to the influence of air and light, is constantly absorbing moisture by its roots, and transmitting it through the branches to the leaves. Here it is partially converted into proper nourishment for the tree, and either added to the substance of the leaves themselves, or returned to the branches; but far the greater portion passes into the air in the form of an invisible vapour. Water cannot be converted into vapour without being combined with heat.* This heat, whether it be supplied direct from the sun or from the leaves, is lost to the tree; consequently, the latter, as well as the surrounding air, is kept cool. When the sun is brightest the evaporation is greatest, the supply from the roots being proportioned to the drain on the leaves.†

* This fact may be familiarly illustrated by moistening the hand with any fluid which evaporates quickly, as ether, spirits of wine, lavender, &c., when a sensation of cold is produced, the hand parting with its heat in order that the liquid may be converted into vapour.

† So, in summer, we choose to walk on the grassy turf, which

A fourth cause of the coolness of the air in a wood, dependent on those already mentioned, may, and probably does, also exist, though I have never had an opportunity of testing the efficacy of this. On the outside of the wood, the air nearest to the ground would naturally be the warmest, and would consequently rise into the upper regions. Colder air from the wood would rush out to supply the place of that which had ascended, and thus a light breeze would be produced, constantly setting from the centre of the grove to its circumference. On this theory we are to account for the superior stillness of our summer nights over the days, the temperature during the absence of the sun being more nearly equalised.

This may seem a long and uncalled for digression; but I am unwilling to pass by any opportunity of drawing the attention of my readers to those instances of design on the part of our Heavenly Father, which, though mainly instrumental to the production of other effects, are greatly conducive to the comfort and enjoyment of mankind. In the present instance we have a striking example of several of the more secret operations of Nature, each exercising a peculiar influence of its own, yet harmoniously combining to produce an effect appreciable by us, and contributing to our advantage. Were we to look on them in another aspect, we should find each cause combining with others to further some

skirts the highway road, rather than on the road itself, the combined effects of radiation and evaporation rendering the grass cooler to the feet. In a winter's morning we prefer the dry road, radiation of heat from the blades of grass having reduced them to so low a temperature, that they are covered with dew or hoar-frost condensed from the warmer atmosphere around them.

end distinct from the first, but, regarded in whatever light we please, co-operating to perfect the will of God:

"Thus all things have their end, yet none but Thine."

For a graphic description of a Beech-wood in autumn, I must refer to Gilpin, who, in his account of Boldrewood, in the New Forest, finds himself compelled to qualify his own strictures on the deficiency of picturesque beauty in this tree. After repeating the substance of the remarks quoted above, he proceeds to say: "If the trees, however, as individuals, were less pleasing, their combinations were highly beautiful, and exhibited much scenery from those natural openings and glades, which are so often found in the internal parts of forests.

"All the woods around Boldrewood Lodge abound in Beech. The mast of this tree is the most fattening food for deer, and gives such repute to the winter venison of Boldrewood walk, that a stranger would have difficulty in getting a king's warrant for a doe executed in it. These woods also afford excellent feeding for hogs, which are led in the autumn season into many parts of the forest to fatten on mast. It is among the rights of the forest borderers to feed their hogs in the forest during pawnage month, as it is called, which commences about the end of September, and lasts six weeks. For this privilege they pay a trifling acknowledgment at the Steward's court at Lyndhurst. The word *pawnage* was the old term for the money thus collected. The method of treating hogs at this season of migration, and of reducing a large herd of these unmanageable brutes to perfect obedience and good government, is curious. The first step the swineherd takes, is to investigate some close, sheltered part of the forest, where there is a conveniency of water, and plenty of Oak or Beech mast. He fixes next on some spreading tree, round the bole* of which he wattles a slight circular fence of the dimensions he wants, and, covering it roughly with boughs and sods, he fills it plentifully with straw or fern.

"Having made this preparation, he collects his colony among the farmers, with whom he commonly agrees for a shilling a head, and will get together perhaps a herd of five or six hundred hogs. Having driven them to their destined habitation, he gives them a plentiful supper of acorns or Beech-mast, which he had already provided, sounding his horn during the repast. He then turns them into the litter, where, after a long journey and a hearty meal, they sleep deliciously. The next morning he lets them look a little around them; shows them the pool, or stream, where they may occasionally drink—leaves them to pick up the offals of the last night's meal—and, as the evening draws on, gives them another plentiful repast under the neighbouring trees, which rain acorns upon them for an hour together at the sound of his horn. He then sends them again to sleep.

"The following day he is probably at the pains of procuring them another meal, with music

* The *bole* or *boll* of a tree; the body of a tree, as a Thorn-boll. The term *bolling trees* is applied to pollards whose heads and branches are cut off, and only the bodies left. So in Exodus, ix. 31, "The flax was bolled," that is, had shot up into a stem.

playing as usual. He then leaves them a little more to themselves, having an eye, however, on their evening hours. But, as they are not hungry, they seldom wander far from home, retiring very early and orderly to bed. After this he throws his sty open, and leaves them to cater for themselves; and from henceforward has little more trouble with them during the whole time of their migration. Now and then, in calm weather, when mast falls sparingly, he calls them perhaps together by the music of his horn to a gratuitous meal; but, in general, they need little attention, returning regularly home at night, though they often wander in the day two or three miles from their sty. There are experienced leaders in all herds, which have spent this roving life before, and can instruct their juniors in the method of it.

"In these forest migrations, it is commonly observed, that, of whatever number the herd consists, they generally separate, in their daily excursions, into such little knots and societies as have formerly had habits of intimacy together; and in these friendly groups they range the forest, returning home at night in different parties, some earlier and some later, as they have been more or less fortunate in the pursuits of the day.

"Besides the hogs thus led out in the mast season to fatten, there are others, the property of forest keepers, which spend the whole year in such societies. After the mast season is over, the indigenous forest hog depends chiefly for his livelihood on the roots of fern; and he would find this food very nourishing, if he could have it in abundance. But he is obliged to procure it by so laborious an operation, that his meals are rarely accompanied with satiety. He continues, however, by great industry, to obtain a tolerable subsistence through the winter, except in frosty weather, when the ground resists his delving snout: then he must perish if he do not in some degree experience his master's care. As spring advances, fresh grasses, and salads of different kinds, add a variety to his bill of fare; and, as summer comes on, he finds juicy berries and grateful seeds, on which he lives plentifully till autumn returns and brings with it the extreme of abundance."*

The Beech tree possesses little legendary interest, and its medicinal virtues, which in Pliny's time were considered numerous, are fallen into disrepute. At Domremy, in Lorraine, formerly stood a Beech tree, under which Joan of Arc, who was born at that place, was supposed to have had her interview with Saint Margaret and Saint Catharine. Another legend is connected with the Beech wood of Saint Leonard, near Horsham. That saint, it is said, wished to rest beneath the Beech trees, but being disturbed during the day by the biting of vipers, and at night by the warbling of nightingales, at his request these animals were removed; since which time, tradition says of the forest,

> "The viper has ne'er been known to sting,
> Or the nightingale e'er heard to sing."

The name Beech is of northern origin; *bece* being the Saxon, *bak* the Swedish and Russian, and *buche* the German name. Its mast was formerly called *buck* in this country. "In some

* Forest Scenery.

parts of France," says Evelyn, "they grind the buck in mills." Buck-wheat, the seed of *Polygonum Fagopyrus*, derives its name from its similarity in shape to the mast of the Beech. The wood of the tree having been formerly used for forming the sides of volumes, the word "book" came to be applied to the volume itself.* The common Beech is always raised from seed, and the varieties are propagated by grafting or budding.

The mast soon loses its germinating power, and is therefore never sown later than the spring of the year which follows its ripening. The seed leaves, which appear above the ground in April or May, are singularly pale, and at the first glance might be mistaken for a fungus. In ten years the tree reaches a height of about twenty feet. In sixty or eighty years it has usually attained its perfection as timber, but lives for a much longer period. It is not well adapted for coppice-wood, ceasing to send up shoots after about thirty or forty years; though if cut down before this

* It is worth noticing how many words connected with literature bear allusion to the materials anciently used in writing, &c. The substances first employed were tables of stone and metal; from this source we derive the expression "Tables of Weights and Measures." Tables of wood were afterwards employed, covered with wax, which were written on by means of an instrument pointed at one end for forming the letters, rounded at the other for the convenience of erasing. this was called a *style*, a word which we retain with an altered meaning. Paper is derived from the Egyptian *papyrus:* we still speak of the leaves of a book, though the *leaves* of the Palm tree are no longer used for the purpose of writing on. *Folio* is from the Latin, *folium*, a leaf. *Liber*, the Latin for a book, meant originally the inner bark of such trees as the Lime, the Ash, the Maple, the Elm, at one period a common writing material: hence we call a collection of books, a *library*. This substance being rolled for the convenience of carriage, a collection of writings was called a *volume*, a name afterwards given to like rolls of paper and parchment.

time, the trees push up again, and the leaves on the shoots so produced seldom fail to remain on the branches during the winter. Young trees generally are, as it has been observed above, liable to the same peculiarity, but not all in the same degree. On this account, fences of young Beech trees may be employed with advantage in flower-gardens, as with their persistent foliage they screen the tender plants during the winter. Gilbert White remarks, that Beeches love to grow in crowded situations, and will insinuate themselves through the thickest covert, so as to surmount it all; they are therefore properly applied to mend thin places in tall hedges: care should be taken, however, not to plant them in situations where the drip might be injurious to the vegetation beneath. Where squirrels are abundant, it is sometimes found necessary to protect the trunks of young Beeches by the application of tar and grease, these destructive little animals being given, especially in spring, to tearing off the bark in strips, in search of the tender inner bark.

An interesting fact recorded by Evelyn* would tend to show that many of our natural Beech woods stand where Oaks originally grew: "That which I would observe to you from the wood at Wooton is, that where goodly Oaks grew, and were cut down by my grandfather almost a hundred years since, is now altogether Beech; and where my brother has extirpated the Beech, there rises Birch. Under the Beech spring up innumerable Hollies, which, growing thick and close together in one of the woods next the meadow, is a *viretum*†

* Letter in Aubrey's "Surrey."

† A leafy wood.

all the year long, which is a very beautiful sight when the leaves of the taller trees are fallen." Strutt also observes that the Beech is of that encroaching and dominant nature, that a wood which may have been originally in equal proportions of Oak and Beech, will in course of time become entirely Beeches.

The leaves of the Beech may be applied to a very useful purpose, even after they have ceased to afford their summer's shelter. Evelyn says, that, "being gathered about the fall, and somewhat before they are much frost-bitten, they afford the best and easiest mattresses in the world to lay under our quilts instead of straw; because, besides their tenderness and loose lying together, they continue sweet for seven or eight years, long before which time straw becomes musty and hard. They are often thus used by divers persons in Dauphiné; and in Switzerland I have sometimes lain on them to my great refreshment." Modern travellers state that in those countries they are still applied to the same purpose.

The nuts of the Beech are rarely used in England except for fattening swine and poultry; but in France an excellent oil is manufactured from them, which is extensively employed both for culinary purposes and for burning; in Silesia it is used by the country people instead of butter. A similar application of Beech mast has been projected in England, but appears never to have been carried into effect. A certain speculator in the reign of George the First proposed a scheme for paying off the national debt with the oil of Beech nuts!

The green wood is heavier than that of any of our

timber trees, but loses nearly a fourth of its weight in drying. Though tolerably hard, it is easily worked, and is applied to a great variety of uses. The principal objection to it is, that it is liable to be perforated by a small beetle. In Scotland, Loudon informs us, the branches and spray are distilled for producing pyroligneous acid; and the wood, branches, and twigs are much used for smoking herrings. It will bear being cut into very thin plates, and is consequently much used for making the scabbards of swords. In Evelyn's time, the art of cutting the wood into these thin plates was not known in England, and when discovered was long kept secret. The neat-looking, but very inconvenient, basket for holding strawberries, called a pottle, is made of Beech. The same material was employed in the days of Evelyn, who refers the custom to remote antiquity. It is also preferred to every other wood for making the wooden shoes called "sabots," worn by the French peasantry. By being dried in the smoke of burning green wood, these acquire the property of resisting the attacks of insects. It forms an excellent fuel, and is no less useful, when converted into charcoal, for the manufacture of gunpowder.

Beechen furniture has been made by poets, both ancient and modern, the emblem of humble rustic content:—

"No wars did men molest
When only Beechen bowls were in request." *

TIBULLUS.

---

* "Nec bella fuerunt,
Faginus adstabat cum scyphus ante dapes."

"If thou, without a sigh, or golden wish,
Canst look upon thy Beechen bowl and dish,—
If in thy mind such power and greatness be,—
The Persian king's a slave, compared with thee.

"Hence, in the world's best years, the humble shed
Was happily and fully furnished:
Beech made their chests, their beds, and their join'd stools;
Beech made the board, the platters, and the bowls."

COWLEY.

"Let herbs to them a bloodless banquet give,
In Beechen goblets let their bev'rage shine,
Cool from the crystal spring their sober wine."

MILTON.

"A Beechen bowl,
A Maple dish, my furniture should be,
Crisp, yellow leaves my bed."

WORDSWORTH.

Several singular varieties of the British Beech are in cultivation, which deserve a passing notice. The Purple Beech has its leaves in their early stage of a bright rose-colour, which, as the season advances, deepens to a rich purple, approaching black. It is a native of Germany, where it was discovered about the middle of the last century. It is usually propagated by grafts, plants raised from seed having a tendency to revert to the common form of the tree. This variety presents a beautiful appearance, when scantily interspersed among other trees in a lawn or grove, but should never be planted alone. The Cut-leaved Beech has its leaves indented, so as almost to resemble in shape the leaves of a fern. The Weeping Beech is said to be the most elegant tree of British growth. A writer in the *Gardener's Magazine* (vol. vii. p. 375) states, that, in the park of J. C. Mountray, county of Tyrone, Ireland, there are some, the trunks of which measure upwards of ten feet in circumference, and that the branches, which extend fifty feet from the stem, touch the ground.

Comparatively few insects attack the Beech, and those which do are chiefly the grubs of moths. The fungi which attack the leaves and bark are more numerous. Among those which grow on the ground in Beech woods, the most

MORELS.

remarkable are the Morel * and the Truffle.† The former of these is a mushroom-like fungus,

* *Morchella esculenta.* † *Tuber cibarium.*

growing in great abundance in the woods of Germany and France, particularly after any of the trees have been burnt down. This fact having been observed led in Germany to the burning of the woods, in order to procure Morels; and, consequently, great numbers of trees were destroyed, till the practice was forbidden by law. They are highly prized for the table, both in their fresh and dry states. In the countries where they abound, many persons gain their livelihood by gathering and drying Morels, which last they effect by running a thread through their

TRUFFLES.

stalks, and hanging them in an airy place. In England they are comparatively rare; but Mr. Berkeley states that he has known them to be so abundant in Kent, as to be used for making a sort of catsup. The Truffle, which is also highly prized in cookery, is very difficult to find, being at all stages of its growth buried beneath the ground. It is black and warty; white within, and marbled with dark veins. "It possesses a strong but agreeable smell, and is generally found by dogs and pigs trained to search for it; but, in those countries where Truffles abound, in the month of October (which is their season for ripening), all the inhabitants repair to the woods, slightly stirring, or rather scratching the ground in those places which experience points out to them as the most likely to contain the tubers. The high price of, and constant demand for Truffles, both in France and other countries, renders this a very lucrative employment; and experienced hunters are rarely deceived in the places where they make their search."* Berkeley (*Eng. Flora,* vol. v. part ii. p. 228) quotes an instance of a poor crippled boy who could detect Truffles with a certainty superior even to that of the best dogs, and so earned a livelihood.

Edible fungi are not peculiar to the Beech woods of Europe. Darwin, in the narrative quoted above, says, "There is one vegetable production deserving notice, from its importance as an article of food to the Fuegians. It is a globular, bright yellow fungus, which grows in vast numbers on the Beech trees. When young it is elastic and turgid, with a smooth surface; but when mature, it shrinks, becomes tougher, and has its entire surface deeply pitted or honey-combed. This has been named by Mr. Berkeley '*Cittaria Darwinii.*' I found a second species on another species of Beech in Chili, and

* Loudon's "Arboretum Britannicum."

a third species has lately been discovered on a third species of Beech in Van Diemen's Land. How singular is this relationship between parasitical fungi and the trees on which they grow, in distant parts of the world! In Tierra del

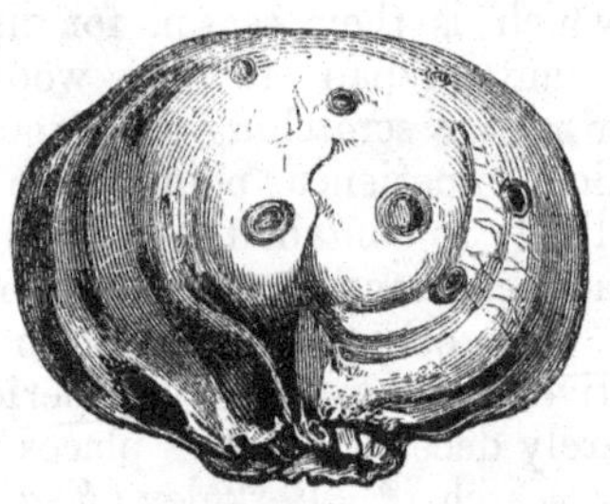

CITTARIA DARWINII.

Fuego, the fungus in its tough and mature state is collected in large quantities by the women and children, and is eaten uncooked. It has a mucilaginous, slightly sweet taste, with a faint smell like that of a mushroom. With the exception of a few berries, chiefly of a dwarf Arbutus, the natives eat no vegetable food besides the fungus. In New Zealand, before the introduction of the potato, the roots of the fern were largely consumed. At the present time, I believe, Tierra del Fuego is the only country in the world where any plant belonging to either of these tribes affords a staple article of food."

Among the many remarkable Beeches now standing in England, the following are most worthy of note.

Of the Beeches in Windsor Park, Jesse says, in

his *Gleanings*, "It is impossible to view these 'sires of the forest,' without feeling a mixture of admiration and wonder. The size of some of them is enormous; one near Sawyer's Lodge measures, at six feet from the ground, thirty-six feet round. It is now protected from injury, and Nature seems to be doing her best towards repairing the damage which its exposure to the attacks of man and beast have produced. It must once have been almost hollow, but the vacuum has been nearly filled up. One might almost fancy that liquid wood, which had afterwards hardened, had been poured into the tree. The twistings and distortions of this huge mass have a curious and striking effect. There is no bark on this extraneous substance, but the surface is smooth, hard, and without any appearance of decay."

In Buckinghamshire, a county which is indebted to this tree for its name,* (Buchen-Heim, the home or land of Beeches,) stand the Burnham Beeches, immortalised by the poet Gray, who gives the following description of them to Horace Walpole:—"I have, at the distance of half a mile, through a green lane, a forest (the vulgar call it a common) all my own, at least as good as so, for I spy no human thing in it but myself. It is a little chaos of mountains and precipices; moun-

* "Here," says Strutt, "the Beech reigns in undivided sovereignty, scarcely admitting an Oak to share its domain, so that we may easily imagine how it must have overrun the country before the opposing influence of agriculture was known; indeed, we are told by old historians, the country was rendered impassable by the thickness of its woods, and the shelter they afforded for marauders and thieves, until several of them were cut down by Leofstan, Abbot of St. Albans." (*Deliciæ Sylvarum.*)

tains, it is true, that do not ascend much above the clouds, nor are the declivities quite so amazing as Dover Cliff; but just such hills as people, who love their necks as much as I do may venture to climb, and crags that give the eye as much pleasure as if they were more dangerous. Both vale and hill are covered with most venerable Beeches, and other very reverend vegetables, that, like most other ancient people, are always dreaming out their old stories to the winds:

> 'And as they bow their hoary tops, relate,
> In murmuring sounds, the dark decrees of Fate:
> While visions, as poetic eyes avow,
> Cling to each leaf, and swarm on every bough.'

At the foot of these I lay myself down, and there grow to the trunk for a whole morning. The timorous hare and sportive squirrel gambol around me, like Adam in Paradise." "The youth to fortune and to fame unknown" has thus enshrined the substance of these remarks, in the most beautiful ode in the English language:—

> "There at the foot of yonder nodding Beech,
> That wreaths its old fantastic roots so high,
> His listless length at noon-tide would he stretch,
> And pore upon the brook that bubbles by."

The Purley Beeches, a beautifully executed engraving of which stands at the beginning of this chapter, are of great antiquity. They appear to be not much known, but their history would be well worth enquiring into. Popular tradition assigns them to the age of William the Conqueror.

Numerous other remarkable trees are noticed by Loudon, Lauder, &c., averaging from twelve to thirty feet in circumference, and from eighty to one hundred and fourteen feet in height. Several figures are also given by Loudon, shewing the tendency of the branches of the Beech to grow together when they touch in crossing. The annexed figure, taken from the *Arboretum Britannicum*, represents a very singular example of this peculiarity in a tree standing in West Hey Wood, between Cliff and Stamford.

BEECH TREE IN WEST HEY WOOD.

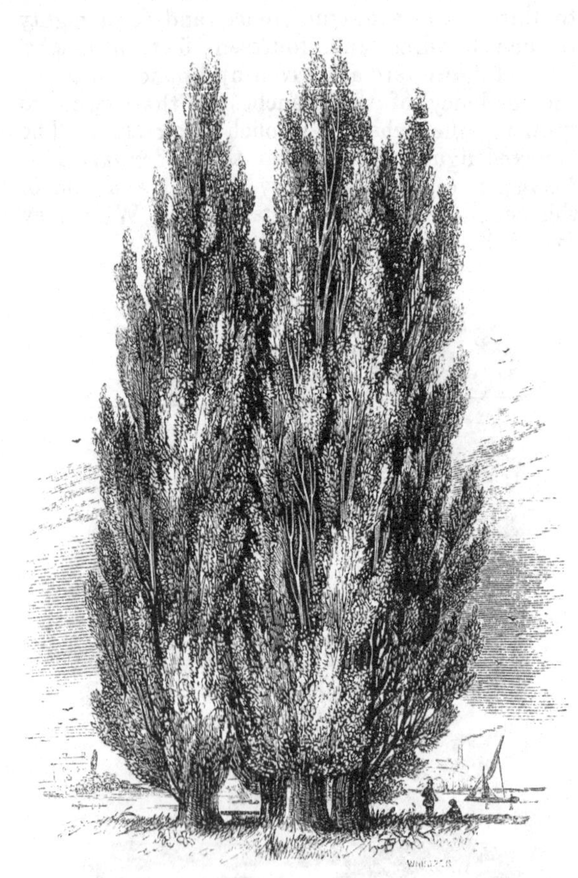

LOMBARDY POPLAR.

THE POPLAR.

Populus.

Class—Diœcia. *Order*—Octandria.

Natural order—Amentaceæ.

No greater contrast can be well imagined than that afforded by the trees of this tribe, when compared with the one which forms the subject of the last chapter. The terms ancient, umbrageous, wide-spreading, picturesque, may be applied to the Beech with propriety; the very reverse of all these will characterise some one or other of the Poplars. The contrast extends even to their places of growth; for while the hill-side is the favourite haunt of the Beech, the Poplar, for the most part, prefers the river's bank. The foliage of the Beech, again, is heavy, unless examined in detail; that of the Poplar scanty, and remarkable for being nearly always in motion, a peculiarity to be attributed to the slenderness and singular formation of its leaf-stalks. It bears its flowers in catkins: these are of two kinds, each growing on separate trees, the barren conspicuous for their length and the large size of the anthers, on which account they have been compared to large red caterpillars; the fertile ones, which are often equally long, may readily be distinguished by the downy wool which invests the seeds, and which is so like cotton, that it has, though with indifferent success, been manufactured into cloth and paper. Most of the tribe are very prolific in suckers from the root. The wood is soft and light, and of little worth in the arts and manufactures. It certainly possesses one property which makes it valuable for some purposes, that, namely, of being very difficult to ignite; hence it may be employed with advantage in flooring rooms. The name *Populus* is said by some to be derived from a Greek word (παιπαλλω, to vibrate) bearing allusion to the tremulous motion of its leaves. Others say that the tree derived its name from being considered at Rome "the tree of the people" (*arbor populi*), a circumstance which brought it into notoriety during the French Revolution.

Four species are indigenous to Great Britain, and many others have been introduced, and are now extensively cultivated. Of these last a short notice will suffice, although one of them, the tall Lombardy Poplar, is probably more generally known than any of the native kinds.

THE WHITE POPLAR, OR ABELE TREE.

POPULUS ALBA.

THE GREY POPLAR.

POPULUS CANESCENS.

THERE appears to be some doubt among authors whether both these trees ought to be considered as natives of Britain, or whether the latter only is indigenous. Evelyn describes the White Poplar, and mentions also a finer sort, "which the Dutch call Abele,* and we have of late much of it transported out of Holland." About the middle of the sixteenth century, as many as 10,000 trees of the same kind are said by Hartlif to have been imported from Flanders, and transplanted into many countries. The fact is, the trees are so much alike in character, that we may safely conclude that the tree which we call the Grey Poplar was known to the earlier writers as a native tree by the name White Poplar, which title was subsequently transferred, for the sake of distinction, to the Abele; the British tree receiving the epithet of "grey" for the same reason. The mere casual observer would scarcely observe the difference between the two; botanists, indeed, are not agreed whether they are

* The English name of Abele is derived from the Dutch name of the tree, Abeel; and this name is supposed by some to be taken from that of the city of Arbela, in the plains of Nineveh, near which, on the banks of the Tigris and Euphrates, great numbers of these trees grow.—*Loudon.*

distinct species, or only varieties. It is, therefore, scarcely worth inquiring to which kind should be referred Cowper's

> "Poplar, that with silver lines his leaf;"

or what tree Barry Cornwall commemorated, when he sung

> "The green woods moved, and the light poplar shook
> Its silver pyramid of leaves."

LEAF OF WHITE POPLAR.

The leaves of both may be distinguished from the other British species by being deeply jagged, the grey less so than the white. The leaves of both are white with down beneath, particularly the latter, which also are larger than those of the other. The fertile catkins of the Abele are oval, and each flower is furnished with four pistils: those of the Grey Poplar are long and cylindrical, and the flowers contain eight pistils each. In all other respects the trees are so similar, that for the remainder of the chapter I shall include them under the same name.

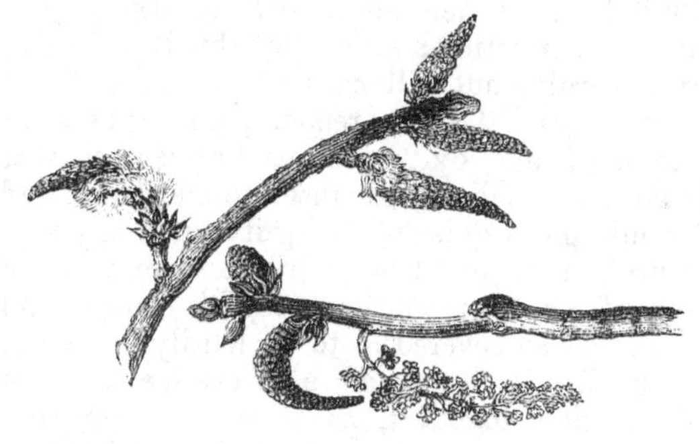

CATKINS OF GREY POPLAR.

The White Poplar was famous in the mythology of the ancients, being consecrated to Hercules, who, in commemoration of one of his victories, gained in a place where this tree was growing in abundance, used to wear a chaplet of its leaves, a custom which was adopted by persons who sacrificed to him. Pliny gravely states that the White Poplar, as well as several other trees which he mentions, always turned its leaves to an opposite quarter of the heavens immediately that the summer solstice was past. Though modern science has not confirmed this observation, the tree may frequently be noticed turning up the white surface of its leaves during the huffling winds which we often experience in summer, and this is a pretty sure indication of approaching rain.*

* "I think there will be rain," a little girl was overheard to say, "for the *weather tree* is shewing its white lining."

The White Poplar is a tree of very rapid growth, attaining a height of from eighty to a hundred feet. When about fifty or sixty years old it is in perfection; soon after this it begins to decay inwardly, but will continue growing for a century longer. Evelyn recommends it as a fit tree to be planted by "such late builders as seat their houses in naked and unsheltered places, and that would put a guise of antiquity upon any new inclosure; since by these, while a man is on a voyage of no long continuance, his house and lands may be so covered as to be hardly known at his return." In England we rarely see many of them growing together, as they are generally plánted to contrast with trees of darker foliage; but in France they are in some places so abundant as to be the prevailing trees in extensive tracts of country, and their wood, called "white wood," is used as fuel, to the exclusion of almost all other firing.

The timber, which is soft and light, was anciently used for making shields, for which its property of yielding before a blow eminently fitted it. Nails may be driven into it without splitting it; hence it may be used with advantage for packing-cases. Being very light, it is made into the rollers used by linen-drapers; and, on account of its uninflammable properties, more than its extreme whiteness, it is well adapted for flooring rooms.

The White Poplar is propagated either by cuttings, by layers, or by suckers, which rise in great numbers from the roots.

BLACK POPLAR.

THE BLACK POPLAR.

Populus nigra.

The Black Poplar, according to Dr. Hunter, derives its name from a black circle perceptible at the centre of its trunk when felled; Loudon seems to think that it received the name from the dark

LEAF OF BLACK POPLAR.

colour of its bark; but it is far more probable that it was originally so called from its having darker foliage than the White Poplar. It may well be distinguished from the other British species by its pointed and slightly notched leaves, which are smooth on both sides.

It was known to Pliny, who recommends it to be planted as a support for vines, a purpose to which, owing to the scantiness of its foliage, it is well adapted. An ancient fable in Roman mythology relates that Phaëton, having obtained permission from his father to drive the horses of the Sun for a day, became terrified, and that Jupiter, to prevent a general conflagration, hurled him from his chariot into the river Po, where he was drowned. His sisters wandered up and down the banks, inconsolable for his loss, till they were converted into Poplars, and wept amber for tears.

> "Nor must the Heliads' fate in silence pass,
> Whose sorrow first produced the Poplar race:
> Their tears, while at a brother's grave they mourn,
> To golden drops of fragrant amber turn."

The Black Poplar is a tree of very rapid growth, and attains a great size. It is consequently often planted as an ornamental tree, though within the last thirty years its place has been much usurped by foreign species. The bright green colour of its foliage never at rest, and sparkling in the light of the sun, especially after a summer shower, is very pleasing to the eye. The seeds when ripe are invested with thick cottony down, and being carried away by the wind, frequently produce trees in situations where they would be least expected. A writer in the *Gardeners' Magazine* states that the kitchen-garden at Versailles was entirely neglected from the beginning of the French Revolution until 1819; and that, in the interval, the light downy seeds of the Black Poplars and Willows of the neighbouring woods had sprung up from the ground, and from the crevices of the walls, and attained even a timber size. The same author records a similar instance in Moscow, where, in 1814, he saw springing up everywhere, from the ashes of those ruined houses which had not been rebuilt, plants of the native Black Poplar.* Thus, had Moscow been left to itself, that immense city would have become by this time a natural forest.

The timber arrives at perfection in about fifty or sixty years, soon after which it begins to decay. In the arts it is of no great value; and, owing to its lightness and softness, is not much used, except for packing-cases and soles of shoes, &c. In Russia the bark is used in the preparation of morocco leather, and in England for tanning leather. Loudon states that the bark of the old trunk is employed by fishermen for buoying up their nets, and mentions other uses to which various parts of the tree may be applied, but none of these are important.

There are many trees of this species existing in Great Britain which exceed seventy feet in height: one at Bury St. Edmund's is said by Strutt (from whose *Sylva* the engraving at the head of this chapter is taken) to be ninety feet high and fifteen feet in circumference at one yard from the ground. The trunk rises forty-five feet before it divides, and then it throws out a profusion of branches. But the largest on record is one, mentioned in Feldborg's *Denmark*,

* Loudon states this to be the Aspen.

in the south of Zealand, near the school of Herlussholm: it is upwards of a hundred feet high, and its trunk is twenty-two feet in circumference. In 1828 it is stated to have been a majestic tree, in full vigour and without a decayed branch.

THE TREMBLING POPLAR, OR ASPEN.

POPULUS TREMULA.

THE Aspen is described by Pliny under the name of Libyan Poplar, and is said to have a very small dark leaf, in great repute for its galls. It is a native of a very extensive range of country, being found throughout the whole of the south of Europe, Asia Minor, and in Lapland to the Frozen Ocean. It prefers wet soils, but is by no means confined to the low countries; for in Scotland it flourishes at an elevation of 1600 feet above the level of the sea. It derives its English name, Aspen, from the German name of the tree, *Espe*, and may readily be distinguished from the other British species by its round leaves, which are of a dark shining green above, and much paler beneath, though destitute of the downy covering which characterises the White Poplar. The leaf-stalk is remarkably long and slender, and being compressed vertically towards its upper extremity, is too weak to support the leaf in a horizontal position. Consequently, the lightest breeze sets it in motion, and hence originated its name, Trembling Poplar.

This peculiarity has obtained for the Aspen the unenviable distinction of being selected as the poetical emblem of restlessness, inconstancy, and fear.

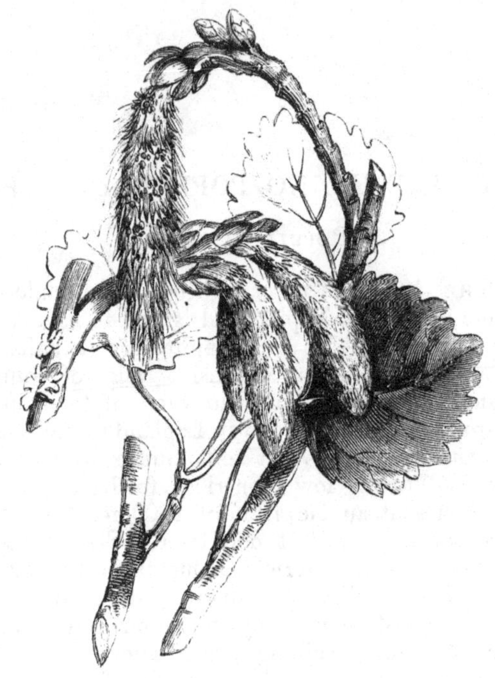

ASPEN.

" With every change his features play'd,
As Aspens shew the light and shade."

" Hearts firm as steel, as marble hard,
'Gainst faith, and love, and pity barr'd,
Have quaked like Aspen leaves in May."

" Variable as the shade,
By the light quivering Aspen made."
SIR W. SCOTT.

" His hand did quake,
And tremble like a leaf of Aspen green."
SPENSER.

On the other hand, an Aspen tree in a state of perfect rest furnishes a beautiful natural emblem of a summer calm. In the following passages, two of our great poets of nature appear to have vied with each other in the selection of their imagery:—

> "Gradual sinks the breeze
> Into a perfect calm, that not a breath
> Is heard to quiver through the closing woods,
> Or rustling turn the many twinkling leaves
> Of Aspen tall. Th' uncurling floods diffused
> In glassy breadth, seem, through delusive lapse,
> Forgetful of their course. 'Tis silence all,
> And pleasing expectation."
>
> THOMSON.

> Into a gradual calm the zephyrs sink,
> A blue rim borders all the lake's still brink;
> And now, on every side, the surface breaks
> Into blue spots, and slowly lengthening streaks.
> Here, plots of sparkling water tremble bright
> With thousand thousand twinkling points of light:
> There, waves, that hardly weltering die away,
> Tip their smooth ridges with a softer ray.
> And now the universal tides repose,
> And, brightly blue, the burnish'd mirror glows,
> Save where, along the shady western marge,
> Coasts with industrious oar the charcoal barge:
> The sails are dropp'd, the Poplar's foliage sleeps,
> And insects clothe, like dust, the glassy deeps.
>
> WORDSWORTH.

Lightfoot tells us that the Highlanders entertain a superstitious notion that our Saviour's cross was made of this tree, for which reason they suppose that its leaves can never rest. Superstitions of this class originated partly in that love of the marvellous which is the characteristic of ignorance, and partly, perhaps, in feelings of real piety; but the sober-minded Christian will not allow his faith to be affected by a mere natural phenomenon. "All is miracle" that tends to confirm his belief in God's superintending Providence, but he humbly refuses to derive from the visible world any teaching but that which Revelation confirms. Reason teaches him that the trembling of the Aspen is dependent on the peculiar mechanism of its leaves, and is to be accounted for by reference to natural causes; and though he fails to discover the purpose of this peculiarity in structure, he is satisfied with observing a new instance of creative power, and prefers to confess his ignorance of design rather than be indebted to nature for evidence which Revelation alone can afford, and which God's Holy Spirit alone can make efficacious.

The Aspen does not generally attain so large a size as the Black Poplar, though there are specimens in existence seventy or eighty feet high. Evelyn says, that "the Aspen thrusts down a more searching foot" than that tree, "and in this likewise differs, that he takes it ill to have his head cut off;" meaning, that the roots extend to a great distance, and that the branches are impatient of pruning. The roots, however, do not descend far beneath the surface, and are remarkable for sending up numerous suckers, which, if the tree be planted in a lawn or garden, are very troublesome, and require to be eaten or mowed down. It is not a long-lived tree, beginning to decay internally when about sixty or eighty years old.

The bark of the Aspen is said to be a favourite food of the beaver, and its leaves are greedily devoured by many domestic, as well as wild animals. The timber is used for nearly the same purposes as that of the other species. As firewood it burns brightly, but rapidly, giving out but little heat.

As an ingredient in the landscape, the Aspen presents the most pleasing appearance in situations where the playful change of its foliage is thrown out by a dark background.

In Belgium it is said to be particularly liable to the attacks of the larvæ of many insects, which are collected by order of the authorities and destroyed.

A chemical principle, called *populine*, has been extracted from the bark and leaves of the Aspen, which has a sweet taste like that of liquorice, and crystallizes in the form of delicate white needles. Its properties are but little known.

FOREIGN POPLARS.

Although the trees belonging to this genus which have been described are undoubtedly indigenous to the soil, the most familiar of all to English readers, and that which is most likely to recur to their minds when "the Poplar" is named, is the tall, gaunt, formal, Lombardy Poplar (*Populus fastigiata*). It is said to be not a native of Europe, but to have been introduced into Pavia from Asia about the year 1590, which would account for the fact that the accurate observer Pliny, who describes the other species, does not mention this. It was introduced into England about the middle of the last century, and some of the original trees stood until within the last ten or twelve years, having attained a height of more than a hundred feet. It is a very fast grower, increasing, when favourably situated, at the unusually rapid rate of five feet in a year, or even more. Some Poplars on the banks of the Seine, near Rouen, had, in 1837, reached the surprising height of 150 feet, having been then planted about thirty-four years.

When Gilpin wrote, the Lombardy Poplar had not long been introduced; he mentions it, however, and points out a peculiarity which it possesses, of being swayed to and fro throughout its whole length by the action of the wind. "Most trees," he says, "in boisterous weather are partially agitated: one side is at rest while the other is in motion. But the Italian Poplar waves in one simple sweep from the top to the bottom, like an ostrich feather on a lady's head. All the branches coincide in the motion; and the least blast makes an impression upon it, when other trees are at rest."

The claims of the Lombardy Poplar to picturesque beauty are very slight. Standing alone, it is rather a deformity than an ornament; nevertheless, planted sparingly in clumps, it has the effect of breaking too level a line, whether formed by round-headed trees in plantations or by large buildings. Loudon devotes seven or eight pages of his *Arboretum* to the proving that in many situations the Lombardy Poplar does possess claims to picturesque beauty; but mentions so many instances of its destroying the harmony of the landscape, that his arguments are far from carrying conviction with them. In England we are most familiar with Poplars planted in a formal row in front of some suburban cottage. In such situations they are sadly misplaced. Amid scenery, too, which has any pretension to wildness, they have a decidedly bad effect, converting the natural woodland into an artificial plantation; for they do not, like most other introduced trees, blend with those by which they are surrounded, but seem to tower above them all, to proclaim, as it were, their foreign origin. This objection, of course, will not apply to them in their native haunts; there, no doubt, they harmonise with the other natural objects around them. I have nowhere seen them so decidedly ornamental as on the banks of the Thames, at Fulham.

The wood is of as little value in the arts as the tree itself in the landscape, being scarcely applied to any other purpose than that of making packing-cases.

The Balsam Poplar, or Tacamahac, (*Populus balsamifera,*) was cultivated in Britain as early as 1692. It is a native of North America, and is remarkable for its large gummy buds, which, as well as its delicately yellow young leaves, diffuse a pleasing odour. The Ontario Poplar (*Populus candicans*) differs from the last in bearing larger leaves, which are heart-shaped at the base.

The necklace-bearing, or Black Italian Poplar, (*Populus monilifera,*) derives the first name from its seed-vessels, which are arranged along a common stem like beads. Why it was first called "Italian" is not so clear, for it appears to have been brought originally from some part of North America. Selby considers it the most valuable of all the Poplars hitherto introduced, as it grows with astonishing rapidity, and produces timber of large size and excellent quality.

The Athenian Poplar (*Populus Græca*) scarcely deserves its classical name, not being a native of Greece, but of the township of Athens in North America. It is a handsome, vigorous tree, approaching nearer to the Aspen than to any other of the Poplars, from which, however, it may be readily distinguished by its longer and more pointed leaves, and the ashen grey colour of the bark on its young branches. It is a very rapid grower, having been observed to produce shoots eight or ten feet long the first season.

Several other kinds of Poplar are cultivated in Great Britain, which resemble more or less those above mentioned, and require no distinct notice.

END OF THE FIRST VOLUME.

LONDON:
Printed by S. & J. Bentley, Wilson, and Fley,
Bangor House, Shoe Lane.

Printed in the United States
By Bookmasters